J H Reed.

ELEMENTS

OF THE

DIFFERENTIAL AND INTEGRAL CALCULUS.

ARRANGED BY

ALBERT E. CHURCH, LL.D.,

PROFESSOR OF MATHEMATICS IN THE U. S. MILITARY ACADEMY.

IMPROVED EDITION,

CONTAINING THE ELEMENTS OF THE

CALCULUS OF VARIATIONS.

NEW YORK:
PUBLISHED BY A. S. BARNES & CO.
51 JOHN STREET.

1855.

PREFACE.

An experience of more than fifteen years, in teaching large Classes in the U. S. Military Academy, has afforded the Author of the following pages unusual opportunities to become familiar with the difficulties encountered by most pupils, in the study of the Differential and Integral Calculus. The results of previous endeavours to remove these difficulties were given to the Public in a former edition. The favour with which that edition has been received, induces him to offer a new one, containing, not only such modifications as have been suggested by a thorough trial in the recitation room, but, in addition, an elementary treatise on the Calculus of Variations. That he has, in some degree, realized the hope of advancing a more general and thorough study of one of the most important auxiliaries to scientific research, is an ample reward for the labour which he has bestowed upon the work.

The Author has in preparation, and expects soon to publish, an Elementary Treatise on Analytical Geometry.

U. S. Military Academy,
West Point, N. Y., August 1, 1850.

CONTENTS.

PART I.

DIFFERENTIAL CALCULUS.

PART II.

INTEGRAL CALCULUS.

PART III.

CALCULUS OF VARIATIONS.

PART I.

DIFFERENTIAL CALCULUS.

FIRST PRINCIPLES.

1. In the branch of Mathematics here treated, as in Analytical Geometry, two kinds of quantities are considered, viz. *variables* and *constants;* the former admitting of an infinite number of values in the same algebraic expression, while the latter admit of but one. The variables are generally designated by the last, and the constants by the first letters of the alphabet.

2. One variable quantity is a function of another, when it is so connected with it, that any change of value in the latter necessarily produces a corresponding change in the former. Thus in the expressions

$$u = bx \qquad\qquad au^2 = cx^3$$

u is a function of x, and x is also a function of u. One of these variables is usually called the function, and the other *the independent variable*, or simply *the variable;* since to one, any arbitrary values may be *assigned*, and from the connection between the two, the corresponding values of the other *deduced.*

This relation is expressed generally thus,

$$u = f(x) \qquad u = \varphi(x) \qquad \text{or} \qquad f(u, x) = 0,$$

f and φ being mere symbols, indicating that u is a function of x. The first two expressions are read, u a function of x, or u equal to a function of x; and the third, a function of u and x equal to zero.

3. Functions are *Increasing* and *Decreasing:*

Increasing, when they are increased if the variable be increased, or decreased if the variable be decreased: *Decreasing* when they are decreased if the variable be increased, or increased if the variable be decreased. In the expressions

$$u = ax^3 \qquad\qquad u = (x + a)^3,$$

u is an increasing function of x. In the expressions

$$y = \frac{1}{x} \qquad\qquad y = (a - x)^3$$

y is a decreasing function of x. In the expression

$$z = (a - y)^2$$

z is a decreasing function for all values of y less than a, but increasing for all values greater than a.

4. Functions are also *Explicit* and *Implicit:*

Explicit, when the value of the function is directly expressed in terms of the variable: *Implicit*, when this value is not directly expressed. In the examples

$$u = (a - x)^2 \qquad\qquad y = \sqrt{a^2 - x^2}$$

u and y are explicit functions of x. In the examples

$$au^2 + bx = cx^2 \qquad y^2 = a^2 - x^2$$

or

$$au^2 + bx - cx^2 = 0 \qquad y^2 + x^2 - a^2 = 0,$$

they are implicit functions of x.

The relation between an implicit function and its variable may be expressed, either by a single equation, as above, or by two or more equations, as

$$u = ay^2 \qquad y^2 = bx,$$

in which u is an implicit function of x. The first relation is indicated generally by

$$f(u, x) = 0,$$

and the other thus,

$$u = f(y) \qquad y = \varphi(x).$$

5. Functions are also *Algebraic* and *Transcendental:*

Algebraic, when the relation between the function and variable can be expressed by the ordinary operations of Algebra, that is, by addition, subtraction, multiplication, division, the formation of powers denoted by constant exponents, and the extraction of roots indicated by constant indices: *Transcendental*, when this relation cannot be so expressed. In the examples

$$u = \log x \qquad u = \sin(a - x) \qquad u = a^x$$

u is a transcendental function of x. If the variable enter any of the exponents, the function is called *Exponential.* If the logarithm of a variable enter, the function is *Logarithmic.* In the expressions

$$u = \sin x \qquad u = \cos x \qquad u = \text{tang}\,\frac{1}{x}$$

u is said to be a *Circular* function.

6. A quantity may be a function of two or more variables, as in the examples

$$u = ax^2 + by \qquad z = axy^2 - ux^2$$

denoted in general thus,

$$u = f(x, y) \qquad z = F(x, y, u).$$

If in a function of a single variable, the latter be made equal to zero, the function reduces to a constant, as in the examples

$$u = ay^2 \qquad u = c + bx^2;$$

if $y = 0$, we have $u = 0$; if $x = 0$, $u = c$.

If in a function of two variables, one be made equal to zero, the function, in general, reduces to a function of the other. So in a function of three variables, if one be made equal to zero, the result will be a function of the other two: If all be zero, the function reduces to a constant; as in the example

$$u = ax + by^2 + cz^3 + d,$$

$z = 0$ gives

$$u = ax + by^2 + d = f(x, y);$$

$z = 0$ and $y = 0$ give

$$u = ax + d = f(x);$$

$z = 0$, $y = 0$, and $x = 0$, give

$$u = d = \text{a constant}.$$

If then in a function of one or more variables, a variable be made equal to zero, the result will be *entirely independent of that variable.* If however in a function of several variables, one be a factor of all the terms containing any of the others; when this variable is 0, the function reduces to a constant, as in the example

$$u = c + ax^2y + bzy^2 = f(x, y, z),$$

$y = 0$ gives

$$u = c.$$

7. To explain what is meant by *the differential of a quantity or function*, let us take the simple expression

$$u = ax^2 \ldots\ldots\ldots\ldots (1)$$

in which u is a function of x. Suppose x to be increased by another variable h; the original function then becomes $a(x + h)^2$; calling this new state of the function u', we have

$$u' = a(x + h)^2 = ax^2 + 2axh + ah^2.$$

From this, subtracting equation (1), member from member, we have

$$u' - u = 2axh + ah^2 \ldots\ldots\ldots\ldots (2).$$

The second member of this equation is the difference between the primitive and new state of the function ax^2, while h is the difference between the two corresponding states of the independent variable x. As the variable h is entirely arbitrary, an infinite number of values may be assigned to it. Let *one* of these values, *which is to remain the same throughout the Calculus*, be denoted by dx, and called *differential of* x, to distinguish it from all other values of h. This particular value being substituted in equation (2), gives for the corresponding difference between the two states of u, or ax^2,

$$u' - u = 2ax.dx + a\,(dx)^2.$$

Now, *the first term of this particular difference is called the differential of* u, and is written

$$du = 2ax.dx.$$

The coefficient $(2ax)$ *of the differential of* x, in this expression, is called, *the differential coefficient of the function* u, and is evidently obtained by dividing the differential of the function by the differential of the variable; and is in general written

$$\frac{du}{dx} = 2ax.$$

Resuming the expression

$$u' - u = 2axh + ah^2,$$

and dividing by h, we have

$$\frac{u' - u}{h} = 2ax + ah.$$

In the first member of this equation, the denominator is the variable increment of the variable x, and the numerator the corresponding increment of the function u; the second member is then the value of the ratio of these two increments. As h is diminished, this value diminishes and becomes nearer and nearer equal to $2ax$, and finally when $h = 0$, it becomes equal to $2ax$. From this we see, that as these increments decrease, their ratio approaches nearer and nearer to the expression $2ax$, and that by giving to h very small values, this ratio may be made to differ from $2ax$, by as small a quantity as we please. This expression is then properly, *the limit of this ratio*, and is at once obtained from the value of the ratio, by making the increment $h = 0$. It will also be seen, that this limit is *precisely the same expression* as the one which we have called the differential coefficient of the function u.

What appears in this particular example is general, for let

$$u = f(x),$$

u being any function of x, and let x be increased by h, then

$$u' = f(x + h).$$

Suppose $f(x + h)$ to be developed, and arranged according to the ascending powers of h, and u to be subtracted from both members, we then have

$$u' - u = Ph + Qh^2 + Rh^3 + \&c. \ldots\ldots\ldots\ldots (3)$$

P, Q, R, &c., being functions of x, and every term of the second member containing h, because $u' - u$ must reduce to 0 when $h = 0$. Substituting for h the particular value dx, and taking the first term for the differential of u, we have

$$du = Pdx, \quad \text{and} \quad \frac{du}{dx} = P.$$

Dividing both members of equation (3) by h, we have

$$\frac{u' - u}{h} = P + Qh + Rh^2 + \&c. \ldots\ldots\ldots\ldots (4).$$

Obtaining the limit of this ratio by making $h = 0$, and denoting it by L, we have

$$L = P,$$

the same value found above for $\frac{du}{dx}$; hence, *the differential coefficient of a function is always equal to the limit of the ratio of the increment of the variable, to the corresponding increment of the function.*

8. The differential of a function of a single variable may then be thus defined. If the variable be increased *by a constant quantity, called the differential of the variable*, and the difference between the new and primitive states of the function be developed according to the ascending powers of the increment; *that term of this difference which contains the first power of the increment is the differential of the function.*

It will in general be found most convenient to obtain first, the differential coefficient, for which we have the following rule:

Give to the variable a variable increment, *find the corresponding state of the function, from which subtract the primitive state, divide the remainder by the increment, obtain the limit of this ratio by making the increment equal to zero,* the result will be the differential coefficient: This, multiplied by the differential of the variable, will give the differential of the function.

The object of the Differential Calculus is, to explain the mode of obtaining and applying the differentials of functions.

9. Let the preceding principles be illustrated by the following

Examples.

1. Let $$u = bx^3.$$

For x put $x + h$, then,

$$u' = b(x + h)^3 = bx^3 + 3bx^2h + 3bxh^2 + bh^3$$

$$u' - u = 3bx^2h + 3bxh^2 + bh^3$$

$$\frac{u' - u}{h} = 3bx^2 + 3bxh + bh^2;$$

passing to the limit, and denoting it by L, we have

$$\mathrm{L} = 3bx^2 = \frac{du}{dx};$$

whence

$$du = 3bx^2dx.$$

2. Let

$$u = ax^2 - cx.$$

Putting $x + h$ for x, and subtracting, we have

$$u' - u = 2axh + ah^2 - ch$$

$$\frac{u' - u}{h} = 2ax + ah - c;$$

making $h = 0$, we have

$$L = 2ax - c = \frac{du}{dx},$$

whence

$$du = 2axdx - cdx.$$

3. Let

$$u = \frac{a}{x},$$

then

$$u' = \frac{a}{x + h}$$

$$u' - u = \frac{a}{x + h} - \frac{a}{x} = \frac{-ah}{x^2 + xh}$$

$$\frac{u' - u}{h} = \frac{-a}{x^2 + xh}$$

and

$$L = -\frac{a}{x^2} = \frac{du}{dx}$$

whence

$$du = -\frac{adx}{x^2}.$$

4. If

$$u = 3ax^3 - mx^4 \qquad du = (9ax^2 - 4mx^3)\,dx.$$

10. Equation (4) article (7) may be put under the form

$$\frac{u' - u}{h} = P + h\,(Q + Rh + \&c.),$$

and if the expression $Q + Rh + \&c.$, (which is a function of x and h,) be represented by P', this becomes

$$\frac{u' - u}{h} = P + P'h \ldots\ldots\ldots\ldots (1);$$

whence

$$u' = u + Ph + P'h^2;$$

that is, the new state of the function *is equal to its primitive state, plus the differential coefficient of the function into the first power of the increment of the variable, plus a function of the variable and its increment, into the second power of the increment.* This expression for the new state of the function being an important one should be carefully remembered.

11. If we resume equation (3) Art. (7), divide by h and transpose P; we have

$$\frac{u' - u}{h} - \text{P} = \text{Q}h + \text{R}h^2 + \&\text{c}.$$

Since when $h = 0$, the expression for the ratio $\frac{u' - u}{h}$ reduces to P, Art. (7); we can plainly assign a value to h so small that

$$\frac{u' - u}{h} < 2\text{P} \qquad \text{or} \qquad \frac{u' - u}{h} - \text{P} < \text{P};$$

whence

$$\text{Q}h + \text{R}h^2 + \&\text{c}. < \text{P},$$

and multiplying by h

$$\text{P}h > \text{Q}h^2 + \text{R}h^3 + \&\text{c}.,$$

which condition will be fulfilled by any value of h which will make $\frac{u' - u}{h} < 2\text{P}$. That is; *in a series arranged according to the ascending powers of a variable, it is always possible to assign to the variable, a value so small as to make the first term numerically greater than the sum of all the others.*

12. If u be an increasing function of x, its new state u' will be greater than u, and

$$\frac{u' - u}{h} = \text{P} + \text{P}'h \ldots\ldots\ldots\ldots \text{Art. (10)}$$

will be positive for all values of h.

If u be a decreasing function, the reverse will be the case, and the ratio be negative for all values of h.

But we see by the preceding article, that when h is sufficiently small, the sum of all the terms that follow P, in the above equation, will be less than P, and therefore the sign of P will be the same as that of the ratio; that is, *positive* when u is an increasing and *negative* when u is a decreasing function. But P is the differential coefficient of u, Art. (7). Hence, *the differential coefficient of an increasing function is always positive; and of a decreasing function, negative.*

It should be observed, that the signs of the differential and differential coefficient are always the same.

13. Let

$$u = v,$$

u and v being functions of the variable x, which are equal to each other for every value of x. If x be increased by h, and u' and v' be the new states of u and v, we have

$$u' = v' \qquad u' - u = v' - v,$$

or placing for u' and v' their values as expressed in Art. (10);

$$Ph + P'h^2 = Qh + Q'h^2,$$

or

$$P + P'h = Q + Q'h,$$

and since P and Q are entirely independent of h, when $h = 0$ there results

$$P = Q \qquad \text{or} \qquad Pdx = Qdx.$$

But P is the differential coefficient of u, and Q the differential coefficient of v, Art. (10), therefore

$$du = dv,$$

that is; *if two functions of the same variable are equal, their differentials will also be equal.*

14. But if $\qquad u = v \pm C,$

u and v being functions of x, and C a constant; and x be increased by h, we have

$$u' = v' \pm C,$$

or placing for v' its value,

$$u' = v + Qh + Q'h^2 \pm C$$

$$\frac{u' - u}{h} = Q + Q'h,$$

and passing to the limit,

$$L = Q = \frac{du}{dx};$$

whence

$$du = Qdx.$$

Q being the differential coefficient of v, Qdx is its differential, therefore

$$du = d\,(v \pm C) = dv,$$

that is; *if two differentials are equal, it does not follow that the expressions from which they were derived, are equal.* We see also, that a constant connected by the sign $\pm$ with a variable, disappears by differentiation. In fact, *the differential of a constant is zero;* since, as it admits of no increase, there is no difference between two states, and of course no differential, Art. (8).

15. Let

$$u = Av,$$

then

$$u' = Av' = A(v + Qh + Q'h^2) \ldots\ldots \text{Art. } (10),$$

$$\frac{u' - u}{h} = A\,(Q + Q'h)$$

$$L = AQ = \frac{du}{dx}; \qquad \text{whence} \quad du = AQdx.$$

But Qdx is the differential of v; therefore

$$du = d\,(Av) = Adv,$$

that is, *the differential of the product of a constant by a variable function, is equal to the constant multiplied by the differential of the function.*

16. When two variable quantities are so connected that one is a function of the other; either may be regarded as the function, and the other as the independent variable. Thus from the expression $u = ax^2$, we readily obtain $x = \sqrt{\frac{u}{a}}$; in which x may be considered a function of the variable u.

In general, let

$$u = f\,(x)\ldots\ldots\ldots\ldots(1);$$

then by deducing the value of x,

$$x = f'\,(u)\ldots\ldots\ldots\ldots(2).$$

In this last expression, let the variable u be increased by any variable increment $u' - u = k$, x will receive the corresponding increment $x' - x$, and the ratio of these increments will be

$$\frac{x' - x}{k}\ldots\ldots\ldots\ldots(3).$$

If the increment $x' - x$ be denoted by h, and we substitute $x + h$ for x, in equation (1), we shall obtain

$$u' - u = Ph + P'h^2 = k,$$

and substituting these values of $x' - x$ and k in expression (3), we have

$$\frac{x'-x}{k}=\frac{h}{\mathrm{P}h+\mathrm{P}'h^2}=\frac{1}{\mathrm{P}+\mathrm{P}'h}.$$

Passing to the limit by making k, the increment of u, equal to 0, in which case $h = 0$, we have

$$\mathrm{L}=\frac{1}{\mathrm{P}}=\frac{dx}{du}.$$

But $\mathrm{P} = \frac{du}{dx}$, hence

$$\frac{dx}{du}=\frac{1}{\frac{du}{dx}},$$

that is, *the differential coefficient of x regarded as a function of u, is the reciprocal of the differential coefficient of u regarded as a function of x.*

To illustrate, take the example

$$u = ax^2;$$

whence

$$x=\sqrt{\frac{u}{a}}.$$

In article (7) we have found $\frac{du}{dx} = 2ax$, then

$$\frac{dx}{du}=\frac{1}{\frac{du}{dx}}=\frac{1}{2ax}=\frac{1}{2a\sqrt{\frac{u}{a}}}=\frac{1}{2\sqrt{au}}$$

17. Let u be an implicit function of x of the second kind, Art. (4), as

$$u = f(y)\ldots\ldots\ldots\ldots(1) \qquad y = \varphi(x)\ldots\ldots\ldots\ldots(2).$$

If x be increased by h, y will receive an increment $y' - y$, which we denote by k; and these increased values of y and x in the second members of (1) and (2) will give

$$u' = u + Qk + Q'k^2 \qquad y' = y + Ph + P'h^2;$$

whence

$$\frac{u' - u}{k} = Q + Q'k \qquad \frac{y' - y}{h} = P + P'h,$$

and by multiplication,

$$\frac{u' - u}{k} \times \frac{y' - y}{h} = QP + Q'Pk + QP'h + \&c.$$

or since $y' - y = k$

$$\frac{u' - u}{h} = QP + Q'Pk + QP'h + \&c.$$

Passing to the limit by making $h = 0$, which gives $k = 0$, we have

$$L = QP = \frac{du}{dx}.$$

But

$$Q = \frac{du}{dy} \qquad \text{and} \qquad P = \frac{dy}{dx};$$

whence

$$\frac{du}{dx} = \frac{du}{dy} \times \frac{dy}{dx},$$

that is, *the differential coefficient of u regarded as a function of x, is equal to the differential coefficient of u regarded as a function of* *y, multiplied by the differential coefficient of y regarded as a function of x.*

If

$$u = f(x)\text{............}(3) \qquad \text{and} \qquad v = \varphi(x)\text{............}(4);$$

in which case u is evidently an implicit function of v; we find from equation (4)

$$x = \varphi'(v)\text{............}(5),$$

and applying the preceding principles to equations (3) and (5), we have

$$\frac{du}{dv} = \frac{du}{dx} \times \frac{dx}{dv}\text{............}(6).$$

But

$$\frac{dx}{dv} = \frac{1}{\frac{dv}{dx}}\text{............Art. (16)},$$

which value in (6) gives

$$\frac{du}{dv} = \frac{\frac{du}{dx}}{\frac{dv}{dx}}.$$

DIFFERENTIATION OF ALGEBRAIC FUNCTIONS.

18. Let

$$u = v \pm w \pm z,$$

in which v w and z are functions of x. Increase x by h, then

$$u' = v' \pm w' \pm z'$$

$$u' - u = (v' - v) \pm (w' - w) \pm (z' - z),$$

from which, after substituting for $(v' - v)$, $(w' - w)$, &c., their values as in article (10), and dividing by h, we have

$$\frac{u' - u}{h} = (\mathrm{Q} + \mathrm{Q}'h) \pm (\mathrm{R} + \mathrm{R}'h) \pm (\mathrm{S} + \mathrm{S}'h).$$

Passing to the limit

$$\mathrm{L} = \mathrm{Q} \pm \mathrm{R} \pm \mathrm{S} = \frac{du}{dx};$$

whence

$$du = \mathrm{Q}dx \pm \mathrm{R}dx \pm \mathrm{S}dx;$$

or since,

$$\mathrm{Q}dx = dv, \qquad \mathrm{R}dx = dw, \qquad \mathrm{S}dx = dz \text{............Art. (8)},$$

$$du = dv \pm dw \pm dz,$$

that is; *the differential of the sum or difference of any number of functions of the same variable is equal to the sum or difference of their differentials taken separately.* Thus, if

$$u = ax^2 - bx^3$$

$$du = d\,(ax^2) - d\,(bx^3) = 2axdx - 3bx^2dx \text{...........Arts. (7 \& 9)}.$$

19. Let uv be the product of any two functions of x, then, if x be increased by h, $u'v'$ will be the new state of the product.

But

$$u' = u + \mathrm{P}h + \mathrm{P}'h^2, \qquad v' = v + \mathrm{Q}h + \mathrm{Q}'h^2,$$

and by multiplication,

$$u'v' = uv + vPh + uQh + PQh^2 + \&c.,$$

thence

$$\frac{u'v' - uv}{h} = vP + uQ + \text{terms containing } h.$$

Passing to the limit we have

$$L = vP + uQ = \frac{d(uv)}{dx};$$

whence

$$duv = vPdx + uQdx = vdu + udv,$$

that is; *The differential of the product of two functions of the same variable, is equal to the sum of the products obtained by multiplying each function by the differential of the other.*

20. Let uvs be the product of three functions. Place

$$uv = r, \qquad \text{then} \qquad uvs = rs,$$

and

$$d(uvs) = d(rs) = rds + sdr \ldots\ldots\ldots\ldots (1).$$

But since

$$r = uv, \qquad dr = udv + vdu;$$

hence by substitution in equation (1)

$$d(uvs) = uvds + sudv + svdu.$$

If we have the product of four functions $uvsw$, we may place $sw = r$, and by a process precisely similar to the above, obtain

$$d(uvsw) = uvsdw + uvwds + uwsdv + vwsdu \ldots\ldots\ldots\ldots (2),$$

and we readily see, that by increasing the number of functions, we may in the same way prove, that *the differential of the pro-*

duct of any number of functions of the same variable, is equal to the sum of the products obtained by multiplying the differential of each into all the others.

21. If we divide both members of equation (2) of the preceding article by $uvsw$, we have

$$\frac{d(uvsw)}{uvsw} = \frac{dw}{w} + \frac{ds}{s} + \frac{dv}{v} + \frac{du}{u},$$

and we should have a similar result for any number of functions; whence we may conclude in general, that *the differential of the product of any number of functions divided by the product, is equal to the sum of the quotients obtained by dividing the differential of each function by the function itself.*

22. Let $$u = v^m,$$

v being any function of x, and m any number, entire or fractional, positive or negative. Increase x by h, then

$$u' = v'^m = (v + Qh + Q'h^2)^m \text{..........Art. (10)},$$

or placing in the binomial formula

$$(x + a)^m = x^m + max^{m-1} + \frac{m(m-1)}{1.2} a^2x^{m-2} + \text{\&c.},$$

$$v \text{ for } x, \quad \text{and} \quad (Qh + Q'h^2) \text{ for } a;$$

we have

$$u' = [v + (Qh + Q'h^2)]^m = v^m + m(Qh + Q'h^2)v^{m-1} + \text{\&c.}$$

$$\frac{u' - u}{h} = m(Q + Q'h)v^{m-1} + \text{\&c.},$$

each of the following terms containing h as a factor. Then

$$L = mv^{m-1}Q = \frac{du}{dx};$$

$$du = dv^m = mv^{m-1}Qdx = mv^{m-1}dv............(1),$$

since $Qdx = dv$, Art. (8). That is, to obtain the differential of any power of a function : *Diminish the exponent of the power by unity, and then multiply by the primitive exponent, and by the differential of the function.*

Examples.

1. If $$u = ax^4,$$

then Art. (15)

$$du = a.dx^4 = a.4x^3dx = 4ax^3dx.$$

2. If $$u = bx^{\frac{2}{3}}$$

$$du = \frac{2}{3}\,bx^{\frac{2}{3}-1}dx = \frac{2}{3}\,bx^{-\frac{1}{3}}\,dx = \frac{2bdx}{3\sqrt[3]{x}}.$$

3. If $$u = cx^{-3}$$

$$du = -\ 3cx^{-4}dx = -\ \frac{3cdx}{x^4}.$$

4. If $$u = (ax - x^2)^5$$

$$du = 5(ax - x^2)^4\ d\ (ax - x^2);$$

but

$$d(ax - x^2) = adx - 2xdx............\text{Art. } (18);$$

hence

$$du = 5(ax - x^2)^4\ (a - 2x)\ dx.$$

23. If in equation (1) of the preceding article we make $m = \frac{1}{n}$, we have

$$dv^{\frac{1}{n}} = \frac{1}{n}v^{\frac{1}{n}-1}dv = \frac{1}{n}v^{\frac{1-n}{n}}dv = \frac{dv}{nv^{\frac{n-1}{n}}},$$

or

$$d\sqrt[n]{v} = \frac{dv}{n\sqrt[n]{v^{n-1}}}.$$

If $n = 2$, we have

$$d\sqrt{v} = \frac{dv}{2\sqrt{v}}$$

that is, the differential of a radical of the second degree is equal to, *the differential of the quantity under the radical sign, divided by twice the radical.*

If $n = 3$, we have

$$d\sqrt[3]{v} = \frac{dv}{3\sqrt[3]{v^2}},$$

and in general the differential of a radical of the nth degree is equal to, *the differential of the quantity under the radical sign divided by n times the $(n-1)$th power of the radical.*

Examples.

1. If $$u = \sqrt{ax^3}$$

$$du = \frac{dax^3}{2\sqrt{ax^3}} = \frac{3ax^2dx}{2\sqrt{ax^3}} = \frac{3}{2}\sqrt{ax}.dx.$$

2. If $u = \sqrt[3]{b - x}$ $du = \frac{-dx}{3\sqrt[3]{(b-x)^2}}$

3. Let $u = \sqrt[5]{bx^2}$ 4. Let $u = \sqrt{2ax - x^2}$.

24. Let
$$u = \frac{s}{v} = sv^{-1},$$

s and v being functions of the same variable, then Art. (19)

$$du = v^{-1}ds + sdv^{-1} = v^{-1}ds - sv^{-2}dv,$$

or

$$du = \frac{ds}{v} - \frac{sdv}{v^2};$$

whence by reducing to a common denominator

$$du = d\,\frac{s}{v} = \frac{vds - sdv}{v^2} \text{............}(1),$$

that is, the differential of a fraction is equal to, *the denominator into the differential of the numerator, minus the numerator into the differential of the denominator, divided by the square of the denominator.*

If the denominator be constant, $dv = 0$, and equation (1) becomes

$$du = \frac{vds}{v^2} = \frac{ds}{v}.$$

If the numerator be constant, $ds = 0$, and equation (1) becomes

$$du = -\frac{sdv}{v^2}$$

In this last case it is evident that u is a decreasing function of v and that its differential should be negative, Art. (12).

Examples.

1. If
$$u = \frac{x}{a - x},$$
$$du = \frac{(a-x)\,dx - xd\,(a-x)}{(a-x)^2} = \frac{(a-x)\,dx + xdx}{(a-x)^2} = \frac{adx}{(a-x)^2}$$

2. If
$$u = \frac{ax^4}{b},$$
$$du = \frac{dax^4}{b} = \frac{4ax^3dx}{b}.$$

3. If
$$u = \frac{c}{ax^3},$$
$$du = -\frac{cdax^3}{(ax^3)^2} = -\frac{3cdx}{ax^4}.$$

25. By a proper application of the preceding principles every algebraic function may be differentiated. Let them be applied to the following

Miscellaneous Examples.

1. If
$$u = (a + bx^n)^p,$$
$$du = p(a + bx^n)^{p-1}d(a + bx^n) \text{............Art. (22);}$$

But
$$d(a + bx^n) = nbx^{n-1}dx;$$

hence
$$du = bnp(a + bx^n)^{p-1}x^{n-1}dx.$$

The solution of this example and many others may be simplified by applying the rule of article (17) thus: make

$$a + bx^n = z, \qquad \text{then} \qquad u = z^p,$$

$$\frac{dz}{dx} = nbx^{n-1} \qquad \frac{du}{dz} = pz^{p-1}.$$

whence

$$\frac{du}{dx} = \frac{du}{dz} \times \frac{dz}{dx} = pz^{p-1} \times nbx^{n-1} = bnp(a + bx^n)^{p-1}x^{n-1},$$

and

$$du = bnp\,(a + bx^n)^{p-1}\,x^{n-1}dx.$$

2. If $$u = (1 - x^2)^3,$$

$$du = 3(1-x^2)^2 d\,(1 - x^2) = -\,6\,(1 - x^2)^2 x dx.$$

3. Let

$$u = \frac{ax}{x + \sqrt{a + x^2}}$$

Place

$$y = x + \sqrt{a + x^2}, \qquad \text{then} \qquad u = \frac{ax}{y},$$

$$dy = dx + \frac{xdx}{\sqrt{a + x^2}} \qquad du = \frac{aydx - axdy}{y^2};$$

hence

$$du = \frac{a\left\{ (x + \sqrt{a + x^2})\,dx - x\left(dx + \frac{xdx}{\sqrt{a + x^2}}\right)\right\}}{(x + \sqrt{a + x^2})^2},$$

or after reduction

$$du = \frac{a^2dx}{(x + \sqrt{a + x^2})^2\,\sqrt{a + x^2}}.$$

4. If $u = \frac{(b+x)^2}{x}$, $du = \frac{(x^2 - b^2)dx}{x^2}$.

5. $u = \sqrt[n]{a^m - x^m}$, $du = -\frac{m}{n}x^{m-1}(a^m - x^m)^{\frac{1}{n}-1}dx$.

6. $u = \frac{2}{\sqrt[3]{a - x^2}}$, $du = \frac{4}{3}x(a - x^2)^{-\frac{4}{3}}dx$.

7. $u = \frac{x}{x - \sqrt{1 - x^2}}$ $du = \frac{-dx}{\sqrt{1 - x^2}\,(x - \sqrt{1 - x^2})^2}$.

8. Let $u = (a - \sqrt{bx^2})^3$. 9. Let $u = \frac{x}{(1+x)^n}$.

10. $u = \frac{1 + x^2}{1 - x^2}$ 11. $u = \left(a - \sqrt{b - \frac{c}{x^3}}\right)^4$.

12. $u = \frac{\sqrt{x^2+1} - 1}{\sqrt{x^2+1} + 1}$. 13. $u = \frac{\sqrt{1+x} + \sqrt{1-x}}{\sqrt{1+x} - \sqrt{1-x}}$.

SUCCESSIVE DIFFERENTIATION.

26. It is readily seen from what precedes, that the differential coefficient of a function of a single variable is, in general, a function of the same variable. It may then be differentiated, and its differential coefficient obtained.

Thus in the example

$$u = ax^3 \qquad \frac{du}{dx} = 3ax^2\ldots\ldots(1),$$

$3ax^2$ is a function of x, different from the primitive function.

If we differentiate both members of equation (1), we have

$$d\left(\frac{du}{dx}\right) = 6ax dx :$$

But since dx is a constant, Art. (24),

$$d\left(\frac{du}{dx}\right) = \frac{d\,(du)}{dx} = \frac{d^2u}{dx};$$

the symbol d^2u, (which is read *second differential* of u) being used to indicate that the function u has been differentiated twice, or that the differential of the differential of u, has been taken. Hence

$$\frac{d^2u}{dx} = 6axdx, \qquad \text{or} \qquad \frac{d^2u}{dx^2} = 6ax,$$

in which dx^2 represents the square of dx, and is the same as if written $(dx)^2$.

The expression, $6ax$, being *the differential coefficient of the first differential coefficient*, is called, *the second differential coefficient.*

To make the discussion general, let $u = f(x)$ and p be its differential coefficient, then

$$\frac{du}{dx} = p \ldots\ldots\ldots\ldots (2).$$

Since p is usually a function of x, let it be differentiated and its differential coefficient be denoted by q, then

$$\frac{dp}{dx} = q \ldots\ldots\ldots\ldots (3).$$

In the same way let q be differentiated and its differential coefficient be r, then

$$\frac{dq}{dx} = r \ldots\ldots\ldots\ldots (4).$$

By differentiating equation (2), we have

$$d\left(\frac{du}{dx}\right) = dp, \qquad \text{or} \qquad \frac{d^2u}{dx} = dp,$$

and by the substitution of this value of dp in (3),

$$\frac{\frac{d^2u}{dx}}{dx} = q, \qquad \text{or} \qquad \frac{d^2u}{dx^2} = q \ldots\ldots\ldots\ldots (5),$$

which is the second differential coefficient of the function.

By differentiating (5), we have

$$\frac{d^3u}{dx^2} = dq,$$

and by the substitution of this value of dq in (4),

$$\frac{\frac{d^3u}{dx^2}}{dx} = r, \qquad \text{or} \qquad \frac{d^3u}{dx^3} = r$$

which is *the differential coefficient of the second differential coefficient*, and is called *the third differential coefficient.*

In the same way the fourth, fifth, &c. may be found, each from the preceding, precisely as the first is obtained from the primitive function.

From the differential coefficients, we may at once obtain the corresponding differentials, by multiplying by that power of dx the exponent of which indicates the order of the required differential, thus

$$d^2u = \frac{d^2u}{dx^2} dx^2 = q\, dx^2,$$

$$\cdot \quad \cdot \quad \cdot \quad \cdot \quad \cdot \quad \cdot \quad \cdot \quad \cdot \quad \cdot \quad \cdot \quad \cdot \quad \cdot \quad \cdot$$

$$d^nu = \frac{d^nu}{dx^n} dx^n \qquad \&c.$$

27. Let

$$u = ax^n,$$

n being a positive whole number, then

$$\frac{du}{dx} = nax^{n-1} \qquad \frac{d^2u}{dx^2} = n(n-1)ax^{n-2},$$

$$\frac{d^3u}{dx^3} = n(n-1)\ (n-2)\, ax^{n-3},$$

. . . .

$$\frac{d^nu}{dx^n} = n(n-1)\ (n-2) \ldots\ldots\ldots 2.1.a.$$

Since the last differential coefficient is constant, its differential will be 0, and we have

$$\frac{d^{n+1}u}{dx^{n+1}} = 0.$$

By examining these results it will be seen, that by each differentiation, the exponent of the variable is diminished by unity. When this exponent is entire and positive, it will finally be reduced to 0, and the corresponding differential coefficient be constant. The next in order, as well as all which follow, will then be 0, and there will be a limited number denoted by n. If n be fractional, by the continued subtraction of unity the result can never be 0, but will finally, if the differentiation be continued, become negative; the successive differential coefficients will then always contain x, and there will be an infinite number. So also if n be negative.

MACLAURIN'S THEOREM.

28. The object of this theorem is, *to explain the manner of developing a function of a single variable, into a series arranged according to the ascending powers of the variable with constant coefficients.*

Let $$u = f(x),$$

and let us assume a development of the proposed form,

$$u = \mathrm{B} + \mathrm{C}x + \mathrm{D}x^2 + \mathrm{E}x^3 + \&c. \ldots\ldots (1),$$

in which B, C, D &c. are entirely independent of x, and depend upon the constants which enter into the given function. It is now required to determine such values for the constants, B, C &c. as will cause the assumed development to be a true one, for all values of x. Since these constants are independent of x, they will not change when we make $x = 0$. If then in (1) we suppose $x = 0$, and denote by A what $f(x)$ or u, becomes under this supposition, we have

$$A = B.$$

Differentiating (1), and dividing by dx, we have

$$\frac{du}{dx} = C + 2Dx + 3Ex^2 + \&c. \ldots\ldots (2);$$

making $x = 0$, and denoting by A′ what $\frac{du}{dx}$ reduces to in this case, we have

$$A' = C.$$

Differentiating (2) and dividing by dx, we have

$$\frac{d^2u}{dx^2} = 2D + 2.3Ex + \&c.;$$

making $x = 0$ and denoting by A″ what $\frac{d^2u}{dx^2}$ becomes, we have

$$A'' = 2D; \qquad \text{whence} \qquad D = \frac{A''}{1.2}.$$

In the same way, denoting by A‴, A⁗ &c., what $\frac{d^3u}{dx^3}, \frac{d^4u}{dx^4}$ &c. become when $x = 0$, we shall find

$$E = \frac{A'''}{1.2.3.} \qquad F = \frac{A''''}{1.2.3.4}, \ \&c.$$

Substituting these values in equation (1), we have

$$u = f(x) = \mathrm{A} + \mathrm{A}'x + \mathrm{A}''\frac{x^2}{1.2} \ldots\ldots + \mathrm{A}^{n\prime}\frac{x^n}{1.2.3\ldots n} + \text{\&c}\ldots(3),$$

in which the general term, or the one which has n terms before it, is, what the nth differential coefficient of the function to be developed becomes when the variable is made equal to 0, multiplied by the nth power of the variable, and divided by the product of the consecutive numbers from 1 to n inclusive.

Examples.

1. Let

$$u = (a + x)^m.$$

This, when $x = 0$, reduces to a^m; hence $\mathrm{A} = a^m$.

By differentiation, &c. we obtain

$$\frac{du}{dx} = m(a + x)^{m-1}, \qquad \frac{d^2u}{dx^2} = m(m - 1)(a + x)^{m-2},$$

$$\frac{d^3u}{dx^3} = m(m - 1)(m - 2)(a + x)^{m-3} \text{ \&c.}$$

Making $x = 0$ in each of these differential coefficients, we have

$\mathrm{A}' = ma^{m-1}$, $\mathrm{A}'' = m(m-1)a^{m-2}$, $\mathrm{A}''' = m(m-1)(m-2)a^{m-3}$, &c.

Substituting these values in the formula (3), we have

$$(a + x)^m = a^m + ma^{m-1}x + \frac{m(m - 1)a^{m-2}x^2}{1.2} + \text{\&c.}$$

2. Let

$$u = \frac{a}{b - x} = a(b - x)^{-1}.$$

By differentiation &c., we have

$$\frac{du}{dx} = a\,(b-x)^{-2} = \frac{a}{(b-x)^2}, \quad \frac{d^2u}{dx^2} = 2a\,(b-x)^{-3} = \frac{2a}{(b-x)^3},$$

$$\frac{d^3u}{dx^3} = 2.3a(b-x)^{-4} = \frac{2.3.a}{(b-x)^4} \ldots\ldots\ldots\ldots \&c.$$

Making $x = 0$ in the original function, and in each differential coefficient, we have

$$A = \frac{a}{b}, \quad A' = \frac{a}{b^2}, \quad A'' = \frac{2a}{b^3} \ldots\ldots\ldots\ldots \&c.$$

These values in the formula (3) give

$$\frac{a}{b-x} = \frac{a}{b} + \frac{a}{b^2}x + \frac{a}{b^3}x^2 + \ldots\ldots \frac{a}{b^{n+1}}x^n + \ldots\ldots$$

3. Let $u = \dfrac{1}{1+x}$ 4. Let $u = \dfrac{1}{\sqrt{1-x^2}}$.

5. $u = \dfrac{1+x}{1-x}$ 6. $u = (1+x^2)^{\frac{5}{3}}$.

Whenever the function to be developed contains the second or higher power of the variable, the work will be much abridged by substituting for this power a single variable, then making the development, and in the result resubstituting the power. Thus, in example 6, by putting z for x^2 we have

$$u = (1 + x^2)^{\frac{5}{3}} = (1 + z)^{\frac{5}{3}},$$

which is easily developed according to the ascending powers of z.

29. Functions which become infinite, when the variable on which they depend is made equal to 0; or any of the differential

coefficients of which become infinite, under the same supposition, cannot be developed by Maclaurin's formula, as in such cases, either the first or some succeeding term of the series would be infinite, while the function itself would not be so.

$$u = \log x \qquad u = \cot x \qquad u = ax^{\frac{1}{2}}$$

are examples of such functions. In the first two A, and in the third A′, would be infinite.

TAYLOR'S THEOREM.

30. A quantity is a function of the sum of two variables, when in the algebraic expression for it, a single variable may be substituted for the sum, and the original function thus reduced, *without a change of form*, to a function of the single variable. Thus

$$u' = a(x + y)^n$$

is such a function, for if in the place of $x + y$ we substitute z, the function becomes $u' = az^n$, a function of z of the same form.

$$\log (x - y)$$

is also such a function of the two variables x and $-y$, which, when for $x - y$ we put z, becomes $\log z$.

31. Let

$$u' = f(x + y).$$

For $x + y$ place z, then

$$u' = f(z) \qquad \text{and} \qquad \frac{du'}{dz} = p\text{......}(1),$$

p the differential coefficient, being entirely independent of dz. If now x be regarded as a constant

$$dz = d(x + y) = dy,$$

and equation (1) becomes

$$\frac{du'}{dy} = p.$$

If y in turn be now regarded as constant

$$dz = d(x + y) = dx,$$

and equation (1) becomes

$$\frac{du'}{dx} = p = \frac{du'}{dy}.$$

That is, *if a function of the sum of two variables be differentiated as though one of the variables were constant; and then the same function be differentiated as though the other variable were constant; and the differential coefficients be taken; these two coefficients will be equal.*

To illustrate, let

$$u' = (x + y)^n, \quad \text{then} \quad du' = n(x + y)^{n-1}\, d(x + y),$$

which if y be regarded as constant becomes,

$$du' = n(x + y)^{n-1}dx; \quad \text{whence} \quad \frac{du'}{dx} = n(x + y)^{n-1},$$

and if x be regarded as constant, the same expression becomes

$$du' = n(x + y)^{n-1}dy; \quad \text{whence} \quad \frac{du'}{dy} = n(x + y)^{n-1}.$$

32. The object of Taylor's Theorem is, *to explain the manner of developing a function of the algebraic sum of two variables, into*

a series arranged according to the ascending powers of one of the variables, with coefficients which are functions of the other and dependent also upon the constants which enter the given function.

Let us write a development of the proposed form,

$$u' = f(x+y) = \text{P} + \text{Q}y + \text{R}y^2 + \text{S}y^3 + \&\text{c}......(1),$$

in which P, Q, R &c. independent of y, are functions of x.

It is required to determine values for them, which substituted in equation (1) will make it true for all values of x and y. If we regard x as constant, differentiate both members of equation (1) with respect to y and divide by dy, we obtain

$$\frac{du'}{dy} = \text{Q} + 2\text{R}y + 3\text{S}y^2 + \&\text{c}.$$

If we regard y as a constant, differentiate equation (1) with respect to x and divide by dx, we obtain

$$\frac{du'}{dx} = \frac{d\text{P}}{dx} + \frac{d\text{Q}}{dx}y + \frac{d\text{R}}{dx}y^2 + \&\text{c}.$$

But by the preceding article we have $\frac{du'}{dy} = \frac{du'}{dx}$; therefore

$$\text{Q} + 2\text{R}y + 3\text{S}y^2 + \&\text{c}. = \frac{d\text{P}}{dx} + \frac{d\text{Q}}{dx}y + \frac{d\text{R}}{dx}y^2 + \&\text{c}.;$$

and since the coefficients of the like powers of y in the two members must be equal,

$$\text{Q} = \frac{d\text{P}}{dx}......(2), \quad 2\text{R} = \frac{d\text{Q}}{dx}......(3), \quad 3\text{S} = \frac{d\text{R}}{dx}......(4).$$

If in equation (1) we make $y = 0$; $f(x+y)$ will reduce to a function of x, Art. (6), which we denote by u. Then

$$u = \text{P}.$$

Substituting this value of P in equation (2), we have

$$Q = \frac{du}{dx}.$$

This value of Q in equation (3), gives

$$2R = \frac{d\left(\frac{du}{dx}\right)}{dx} = \frac{d^2u}{dx^2}; \text{ whence } R = \frac{d^2u}{1.2.dx^2};$$

and this value of R in (4) gives

$$3S = \frac{d\left(\frac{d^2u}{1.2.dx^2}\right)}{dx} = \frac{d^3u}{1.2.dx^3}; \text{ whence } S = \frac{d^3u}{1.2.3.dx^3}.$$

By the substitution of these values of P, Q, R &c. in equation (1) we have Taylor's formula;

$$u' = f(x + y) = u + \frac{du}{dx}\frac{y}{1} + \frac{d^2u}{dx^2}\frac{y^2}{1.2} + \ldots\ldots \frac{d^nu}{dx^n}\frac{y^n}{1.2.3\ldots n} + \ldots$$

By an examination of the several terms of this formula, we see that the first (u) is what the function to be developed becomes, when the variable, according to the ascending powers of which the series is to be arranged, is made equal to 0. The second $\left(\frac{du}{dx}\frac{y}{1}\right)$ is the first differential coefficient of the first term, multiplied by the first power of this variable; and the general term is *the* nth *differential coefficient of the first term*, multiplied by the nth power of the variable, and divided by the product of the consecutive numbers from 1 to n inclusive.

The development of $f(x - y)$ is obtained from the formula by changing $+ y$ into $- y$; thus

$$f(x - y) = u - \frac{du}{dx}y + \frac{d^2u}{dx^2}\frac{y^2}{1.2} - \frac{d^3u}{dx^3}\frac{y^3}{1.2.3} + \&c.$$

Examples.

1. Let $$u' = (x + y)^m.$$

Making $y = 0$ we obtain $u = x^m$, and thence by differentiation,

$$\frac{du}{dx} = mx^{m-1}, \qquad \frac{d^2u}{dx^2} = m(m-1)x^{m-2},$$

$$\frac{d^3u}{dx^3} = m(m-1)(m-2)x^{m-3}, \frac{d^nu}{dx^n} = m(m-1)\ldots\ldots(m-n+1)x^{m-n},$$

These values being substituted in the formula give

$$u' = (x+y)^m = x^m + mx^{m-1}y + \frac{m(m-1)x^{m-2}y^2}{1.2} + \ldots\ldots$$

$$\frac{m(m-1)\ldots\ldots(m-n+1)x^{m-n}y^n}{1.2.3\ldots\ldots n} + \ldots\ldots$$

If it were required to develope the function in terms of the ascending powers of x we should make $x = 0$, and obtain for the first term y^m, from which the other terms are derived as before.

2. Let $$u' = \frac{a}{x+y}.$$

Making $y = 0$, we obtain, $u = \frac{a}{x}$ for the first term; thence

$$\frac{du}{dx} = -\frac{a}{x^2}, \qquad \frac{d^2u}{dx^2} = \frac{2a}{x^3},$$

$$\frac{d^3u}{dx^3} = -\frac{2.3.a}{x^4}. \qquad \frac{d^nu}{dx^n} = \pm\frac{2.3\ldots\ldots na}{x^{n+1}}.$$

These values being substituted in the formula give

$$u' = \frac{a}{x+y} = \frac{a}{x} - \frac{a}{x^2}y + \frac{a}{x^3}y^2\ldots\ldots\ldots\ldots \pm \frac{a}{x^{n+1}}y^n\ldots\ldots$$

3. Develope $u' = \frac{b}{(x - y)^{\frac{1}{3}}}$...according to the powers of $-y$.

4. $u' = \frac{a}{(x - y)^2}$... " " of x.

33. Since in the formula of Taylor, the coefficients of the different powers of one variable are functions of the other, it is plain that if such a value be assigned to the other, as to reduce any of these coefficients to infinity, the second member will become infinite, and the formula fail to give a development for this particular value; as, in this case, the first member will become a function of the first variable, which function cannot possibly be equal to infinity for a particular value of the second variable, on which it in no way depends. Thus, in the example

$$u' = \sqrt{a + x + y}$$

which, when developed according to the ascending powers of y, gives

$$u' = \sqrt{a + x} + \frac{1}{2\sqrt{a + x}} y - \frac{1}{8\sqrt{(a + x)^3}} y^2 \ldots\ldots\ldots,$$

the particular value $x = -a$ reduces the coefficients of the powers of y to infinity, while the original function is reduced to $\sqrt{y}$: We should thus have $\sqrt{y} = \infty$, which cannot be. For every other value of x, however, these coefficients will be finite and the development true.

The difference between this failing case and that of Maclaurin's formula is marked. In this, the failure is only for a particular value of that variable which enters the coefficients, all other values of both variables giving a true development; while in the former case, if the formula fails to develope a function for one value of the variable, it fails for every other value.

34. If
$$u = f(x),$$

and x be increased by h, we have for the second state

$$u' = f(x + h),$$

and by changing y into h in Taylor's formula, we obtain

$$u' = f(x + h) = u + \frac{du}{dx} h + \frac{d^2u}{dx^2} \frac{h^2}{1.2} + \&c.;$$

which is *the development of the second state of a function.*

From this we have

$$u' - u = \frac{du}{dx} h + \frac{d^2u}{dx^2} \frac{h^2}{1.2} + \frac{d^3u}{dx^3} \frac{h^3}{1.2.3} + \&c.$$

If we now put for h the particular value dx, we have

$$u' - u = du + \frac{d^2u}{1.2} + \frac{d^3u}{1.2.3} + \&c.$$

35. If in the development of $f(x + y)$ by Taylor's formula, we suppose $x = 0$, and represent by A, A′, A″, &c. what u, $\frac{du}{dx}$, $\frac{d^2u}{dx^2}$, &c. become under this supposition, we have

$$f(y) = A + A'y + \frac{A''y^2}{1.2} + \frac{A'''y^3}{1.2.3} + \&c.$$

A, A′, A″, &c. being constant, and since y is the only variable we may write x for it, and thus have

$$f(x) = \text{A} + \text{A}'x + \frac{\text{A}''x^2}{1.2} + \frac{\text{A}'''x^3}{1.2.3} + \&c.$$

which is identical with Maclaurin's formula.

DIFFERENTIATION OF TRANSCENDENTAL FUNCTIONS.

36. Let us take, first, the exponential function

$$u = a^x,$$

and increase x by h, then,

$$u' = a^{x+h} = a^x a^h \ldots\ldots\ldots\ldots(1).$$

a^h being a function of the single variable h, may be developed into a series, the first term of which [being what a^h becomes when $h = 0$, Art. (28)], is $a^0 = 1$. We may then write

$$a^h = 1 + kh + k'h^2 + k''h^3 + \&c.,$$

k, k', k'', &c. being constants depending upon a. By substituting this value of a^h in (1), we obtain

$$u' = a^x(1 + kh + k'h^2 + k''h^3 + \&c.);$$

whence

$$\frac{u' - u}{h} = a^x k + a^x k'h + \&c.$$

Passing to the limit of this ratio, we have

$$\text{L} = a^x k = \frac{du}{dx}, \quad \text{and} \quad du = a^x k dx \ldots\ldots\ldots\ldots(2).$$

To determine the value of k, let us develope $u = a^x$ by Maclaurin's formula. We have

$$u = a^x, \qquad \frac{du}{dx} = a^x k \cdot$$

$$d\left(\frac{du}{dx}\right) = kda^x = k^2a^xdx; \quad \text{whence} \quad \frac{d^2u}{dx^2} = k^2a^x;$$

$$d\left(\frac{d^2u}{dx^2}\right) = k^2da^x = k^3a^xdx; \text{ whence } \quad \frac{d^3u}{dx^3} = k^3a^x; \text{ \&c.}$$

Making $x = 0$ in these expressions, we find for the coefficients in the formula

$$\text{A} = a^0 = 1, \quad \text{A}' = a^0k = k, \quad \text{A}'' = k^2, \quad \text{A}''' = k^3, \text{ \&c.};$$

whence

$$u = a^x = 1 + kx + \frac{k^2x^2}{1.2} + \frac{k^3x^3}{1.2.3} + \text{\&c.}$$

In this, let $x = \frac{1}{k}$, then

$$a^{\frac{1}{k}} = 1 + 1 + \frac{1}{1.2} + \frac{1}{1.2.3} + \text{\&c.}$$

Summing this series, we have

$$a^{\frac{1}{k}} = 2,7182818\ldots\ldots\ldots\ldots$$

This number is the base of the Naperian system of logarithms, which is usually denoted by e; we then have

$$a^{\frac{1}{k}} = e, \qquad \text{and} \qquad a = e^k;$$

hence k is the Naperian logarithm of a, denoted by la. Then

$$du = da^x = a^xladx,$$

that is, the differential of a constant raised to a power denoted by a variable exponent, is equal to *the power, multiplied by the Naperian logarithm of the root into the differential of the exponent.*

37. Resuming the expressions

$$u = a^x \qquad \frac{du}{dx} = a^x la,$$

regarding u as the independent variable and x as the function, we have, Art. (16),

$$\frac{dx}{du} = \frac{1}{a^x la}; \qquad \text{whence} \qquad dx = \frac{du}{u}\frac{1}{la}.$$

If a be the base of any system of logarithms, then $x = \log u$ in that system, and

$$dx = d \log u = \frac{du}{u}\frac{1}{la}.*$$

---

* NOTE.—Throughout the book, the symbol l before a quantity will indicate the Naperian logarithm of that quantity. Since the logarithms of the same number in different systems are as the moduli, we have

$$\log a : la :: \text{M} : 1,$$

and when a is the base of a system, since $\log a = 1$

$$1 : la :: \text{M} : 1;$$

whence

$$\text{M} = \frac{1}{la}.$$

Also,

$$\log e : le :: \text{M} : 1,$$

and since $le = 1$

$$\text{M} = \log e.$$

The modulus of a system, then, admits of two forms of expression, both of which should be remembered. The one is, *unity divided by the Naperian logarithm of the base of the system whose modulus is required;* the other, *the logarithm of the Naperian base taken in the system whose modulus is required.*

But $\frac{1}{la}$ is the modulus of the system whose base is a; then

$$d \log u = \text{M} \frac{du}{u}.$$

For the Naperian system, M = 1, and this expression becomes

$$dlu = \frac{du}{u}.$$

The differential of the logarithm of a quantity, is then equal to *the modulus of the system into the differential of the quantity divided by the quantity;* and this in the Naperian system, becomes *the differential of the quantity divided by the quantity.*

Examples.

1. If $$u = l(ax^3)$$

$$du = \frac{dax^3}{ax^3} = \frac{3ax^2dx}{ax^3} = 3\frac{dx}{x}$$

2. If $$u = l\left(\frac{a}{a-x}\right)$$

$$du = \frac{d\left(\frac{a}{a-x}\right)}{\frac{a}{a-x}} = \frac{\frac{adx}{(a-x)^2}}{\frac{a}{a-x}} = \frac{dx}{a-x}.$$

Otherwise thus, $u = l\left(\frac{a}{a-x}\right) = la - l(a-x)$

$$du = dla - dl(a-x) = \frac{dx}{a-x}$$

3. Let $$u = l\left(\frac{\sqrt{1+x^2}+x}{\sqrt{1+x^2}-x}\right).$$

Multiply both terms of the fraction by the numerator, then

$$u = l(\sqrt{1+x^2}+x)^2 = 2l(\sqrt{1+x^2}+x),$$

$$du = \frac{2d(\sqrt{1+x^2}+x)}{\sqrt{1+x^2}+x} = \frac{2dx}{\sqrt{1+x^2}}.$$

4. If $u = \dfrac{1}{\sqrt{-1}}\, l(x\sqrt{-1}+\sqrt{1-x^2}),\quad du = \dfrac{dx}{\sqrt{1-x^2}}.$

5. If $u = l\left(\dfrac{\sqrt{1+x}+\sqrt{1-x}}{\sqrt{1+x}-\sqrt{1-x}}\right),\quad du = -\dfrac{dx}{x\sqrt{1-x^2}}.$

6. Let $u = l\left(\dfrac{a-x}{x}\right)^2.$ 7. Let $u = l\,(a+x)^2\,(a-x)^2.$

8. $u = l\left(\dfrac{\sqrt{1-x}}{\sqrt{1+x^2}}\right).$ 9. $u = l(a - x^2)\sqrt{x}).$

10. Let $$u = (lx)^n.$$

Make $lx = z,$ then $u = z^n\,;$

$$du = nz^{n-1}dz = \frac{n\,(lx)^{n-1}\,dx}{x}.$$

11. Let $$u = l(lx).$$

Make $lx = z,$ then $u = lz\,;$

$$du = \frac{dz}{z} = \frac{dx}{xlx}.$$

12. Let $$u = a^{lx}.$$

38. It has been seen, Art. (29), that log x cannot be developed according to the ascending powers of x. To obtain a logarithmic series, let us take $u = \log(1 + x)$ and develope it by Maclaurin's formula. By differentiation &c.

$$du = \frac{Mdx}{1+x}, \qquad \frac{du}{dx} = \frac{M}{1+x} = M(1+x)^{-1};$$

$$\frac{d^2u}{dx^2} = -M(1+x)^{-2} = \frac{-M}{(1+x)^2}; \qquad \frac{d^3u}{dx^3} = \frac{2M}{(1+x)^3}.$$

Making $x = 0$, we have for the values of A, A′, A″ &c., in the formula,

$$A = \log 1 = 0 \qquad A' = M \qquad A'' = -M \qquad A''' = 2M, \&c;$$

whence

$$u = \log(1+x) = M\left(x - \frac{x^2}{2} + \frac{x^3}{3} \ldots\ldots \pm \frac{x^n}{n} \ldots\ldots\right),$$

in which the logarithm of a quantity is expressed by a series, arranged according to the ascending powers of a quantity less by unity.

39. By the aid of logarithms we may simplify the differentiation of complicated exponential functions. For example:

1. Let $$u = z^y,$$

z and y being any functions of the same variable. Take the Naperian logarithms of both members, then

$$lu = lz^y = ylz;$$

and by differentiation

$$\frac{du}{u} = dylz + y\frac{dz}{z};$$

whence

$$du = u(dylz + y\frac{dz}{z}) = z^y lzdy + yz^{y-1}dz,$$

which is evidently *the sum of the differentials, taken by first regarding y as the only variable, and then z.*

2. Let $$u = a^{b^x}.$$

Taking the logarithms of both members

$$lu = b^x la, \qquad \frac{du}{u} = la\,.\,db^x = lab^x lbdx,$$

$$du = a^{b^x}\,b^x lalb\,dx.$$

3. Let. $$u = z^{t^s},$$

then

$$lu = t^s lz, \qquad \frac{du}{u} = \frac{t^s dz}{z} + lz(t^s ltds + st^{s-1}dt),$$

$$du = z^{t^s}t^s(\frac{dz}{z} + lzltds + \frac{s}{t}\,lzdt).$$

DIFFERENTIATION OF THE CIRCULAR FUNCTIONS.

40. Since any arc of a circle, when less than 90°, is greater than its sine, and less than its tangent; we must have for all values of y less than 90°,

$$\frac{\sin y}{y} < 1 \qquad \text{and} \qquad \frac{\sin y}{\operatorname{tang} y} < \frac{\sin y}{y}.$$

But

$$\operatorname{tang} y = \frac{\mathrm{R} \sin y}{\cos y}; \qquad \text{whence} \qquad \frac{\sin y}{\operatorname{tang} y} = \frac{\cos y}{\mathrm{R}} \ldots\ldots(1).$$

Making $y = 0$, $\cos y$ becomes R, and we have for the limit of the ratio (1),

$$\mathrm{L} = \frac{\mathrm{R}}{\mathrm{R}} = 1.$$

and since $\frac{\sin y}{y}$ cannot exceed unity, nor be less than $\frac{\sin y}{\operatorname{tang} y}$, we also have its limit $= 1$; that is, *the limit of the ratio of an arc to its sine is unity.*

41. Let

$$u = \sin x.$$

Increase x by h, then

$$u' = \sin(x + h), \qquad u' - u = \sin(x + h) - \sin x,$$

or by placing $x + h$ for p and x for q in the formula,

$$\sin p - \sin q = \frac{2}{\mathrm{R}}\left[\sin \tfrac{1}{2}(p - q) \cos \tfrac{1}{2}(p + q)\right],$$

$$u' - u = \frac{2}{R} \sin \tfrac{1}{2}h \cos (x + \tfrac{1}{2}h).$$

Dividing both members by h, and then both terms of the fraction in the second member by 2,

$$\frac{u' - u}{h} = \frac{\sin \frac{1}{2}h}{\frac{1}{2}h} \frac{\cos (x + \frac{1}{2}h)}{R},$$

and passing to the limit, since $\left(\dfrac{\sin \frac{1}{2}h}{\frac{1}{2}h}\right)_{h=0} = 1,$*

$$L = \frac{\cos x}{R} = \frac{du}{dx};$$

whence

$$du = d \sin x = \frac{\cos x \, dx}{R}.$$

If $\qquad u = \cos x,$

$$du = d \cos x = d \sin (90^\circ - x) = \frac{\cos (90^\circ - x)}{R} d(90^\circ - x);$$

whence

$$d \cos x = - \frac{\sin x}{R} dx.$$

If $\qquad u = \text{ver-sin } x$

$$du = d \text{ ver-sin } x = d(R - \cos x) = - d \cos x;$$

whence $\qquad d \text{ ver-sin } x = \dfrac{\sin x}{R} dx.$

---

* NOTE.—This notation indicates that the expression for the quantity within the parenthesis becomes unity when $h = 0$.

If $$u = \text{tang}\, x$$

$$du = d\,\text{tang}\, x = d\,\frac{\text{R} \sin x}{\cos x}$$

$$= \text{R}\,\frac{(\cos x d \sin x - \sin x d \cos x)}{\cos^2 x} = \frac{dx(\cos^2 x + \sin^2 x)}{\cos^2 x}.$$

whence

$$d\,\text{tang}\, x = \frac{\text{R}^2 dx}{\cos^2 x}.$$

If $$u = \cot x$$

$$du = d \cot x = d\,\text{tang}\,(90^\circ - x) = \text{R}^2\,\frac{d\,(90^\circ - x)}{\cos^2 (90^\circ - x)};$$

whence

$$d \cot x = -\frac{\text{R}^2 dx}{\sin^2 x}.$$

If $$u = \sec x$$

$$du = d \sec x = d\,\frac{\text{R}^2}{\cos x} = \frac{\text{R} \sin x\, dx}{\cos^2 x};$$

whence

$$d \sec x = \frac{\text{tang}\, x\, dx}{\cos x} = \frac{\text{tang}\, x \sec x}{\text{R}^2}\, dx.$$

If $$u = \text{cosec}\, x$$

$$du = d\,\text{cosec}\, x = d \sec\,(90^\circ - x) = \frac{\cot x . \text{cosec}\, x\; d(90^\circ - x)}{\text{R}^2};$$

whence

$$d \operatorname{cosec} x = -\frac{\cot x . \operatorname{cosec} x \, dx}{R^2}.$$

If $R = 1$, these formulas become

$$d \sin x = \cos x \, dx, \qquad d \cos x = -\sin x \, dx,$$

$$d \text{ ver-sin } x = \sin x \, dx, \qquad d \operatorname{tang} x = \frac{dx}{\cos^2 x}, \text{ \&c.}$$

and should be remembered.

Examples.

1. If $$u = \sin \frac{bx}{a}$$

$$du = \cos \frac{bx}{a} d \frac{bx}{a} = \frac{b}{a} \cos \frac{bx}{a} dx.$$

2. If $$u = \cos \frac{1}{x}$$

$$du = -\sin \frac{1}{x} d \frac{1}{x} = \frac{1}{x^2} \sin \frac{1}{x} dx.$$

3. If $$u = \operatorname{tang} (a - x)^2$$

$$du = \frac{d(a - x)^2}{\cos^2 (a - x)^2} = -\frac{2(a - x)dx}{\cos^2(a - x)^2}.$$

4. If $$u = \cot^2 x$$

$$du = 2 \cot x \, d \cot x = -\frac{2 \cot x \, dx}{\sin^2 x}$$

5. If $$u = (\cos x)^{\sin x},$$

make $\cos x = z$, $\sin x = y$; then $u = z^y$, and Art. (39)

$$du = z^y lz dy + yz^{y-1}dz = dx\,(\cos x)^{\sin x}\left(\cos x\ l\cos x - \frac{\sin^2 x}{\cos x}\right).$$

6. Let $u = \dfrac{\sin(1+x)}{x}$. 7. Let $u = \text{tang}\,(-m\sqrt{x})$.

42. In the preceding article we have found the differentials of the sine, cosine, &c. in terms of the arc as an independent variable; let it now be required to find the differential of the arc, in terms of its sine, cosine, &c.

If $u = \sin x$, then $x = \sin^{-1}u$,*

$$du = \frac{\cos x\,dx}{R}, \quad \text{and} \quad \frac{du}{dx} = \frac{\cos x}{R}.$$

If now x be regarded as the function, and u as the independent variable, we have, Art. (16),

$$\frac{dx}{du} = \frac{1}{\frac{du}{dx}} = \frac{R}{\cos x},$$

and since $\cos x = \sqrt{R^2 - \sin^2 x} = \sqrt{R^2 - u^2}$

$$\frac{dx}{du} = \frac{R}{\sqrt{R^2 - u^2}}; \quad \text{whence} \quad dx = \frac{R\,du}{\sqrt{R^2 - u^2}}$$

---

* NOTE.—The notation $\sin^{-1} u$, $\text{tang}^{-1} u$, &c., is used to designate *the arc whose sine is u; whose tangent is u, &c.*

If

$$u = \cos x, \qquad x = \cos^{-1} u, \qquad \frac{du}{dx} = -\frac{\sin x}{\text{R}};$$

$$\frac{dx}{du} = \frac{-\text{R}}{\sin x} = -\frac{\text{R}}{\sqrt{\text{R}^2 - \cos^2 x}} = -\frac{\text{R}}{\sqrt{\text{R}^2 - u^2}};$$

whence

$$dx = -\frac{\text{R}du}{\sqrt{\text{R}^2 - u^2}}.$$

If

$$u = \text{ver-sin}\, x, \qquad x = \text{ver-sin}^{-1} u;$$

$$\frac{du}{dx} = \frac{\sin x}{\text{R}} \quad \text{and} \quad \frac{dx}{du} = \frac{\text{R}}{\sin x};$$

or since

$$\sin x = \sqrt{(2\text{R} - \text{ver-sin}\, x)\,\text{ver-sin}\, x} = \sqrt{(2\text{R} - u)u};$$

$$\frac{dx}{du} = \frac{\text{R}}{\sqrt{(2\text{R} - u)u}}; \quad \text{whence} \quad dx = \frac{\text{R}du}{\sqrt{2\text{R}u - u^2}}.$$

If

$$u = \text{tang}\, x, \qquad x = \text{tang}^{-1} u, \qquad \frac{du}{dx} = \frac{\text{R}^2}{\cos^2 x};$$

$$\frac{dx}{du} = \frac{\cos^2 x}{\text{R}^2} = \frac{\text{R}^2}{\sec^2 x} = \frac{\text{R}^2}{\text{R}^2 + \text{tang}^2 x};$$

whence

$$dx = \frac{\text{R}^2 du}{\text{R}^2 + u^2}.$$

If R = 1, these formulas become

$$dx = \frac{du}{\sqrt{1-u^2}}. \qquad dx = -\frac{du}{\sqrt{1-u^2}}.$$

$$dx = \frac{du}{\sqrt{2u-u^2}}. \qquad dx = \frac{du}{1+u^2}.$$

Examples.

1. If $$x = \sin^{-1} 2u\sqrt{1-u^2},$$

$$dx = \frac{d(2u\sqrt{1-u^2})}{\sqrt{1-(2u\sqrt{1-u^2})^2}} = \frac{2du}{\sqrt{1-u^2}}.$$

2. If $$x = \text{tang}^{-1}\left(-\frac{c}{y}\right),$$

$$dx = \frac{d\left(-\frac{c}{y}\right)}{1+\left(\frac{c}{y}\right)^2} = \frac{cdy}{c^2+y^2}.$$

3. If $u = \cos^{-1}\frac{y}{a-y}$, $\quad du = \frac{-ady}{(a-y)\sqrt{a^2-2ay}}$.

4. If $u = \text{ver-sin}^{-1}\frac{1}{x}$, $\quad du = \frac{-dx}{x\sqrt{2x-1}}$.

43. We are now able to develope $\sin x$, $\cos x$, &c., in terms of the ascending powers of x, by Maclaurin's formula.

1. If

$$u = \sin x, \qquad \text{and} \qquad R = 1;$$

$$\frac{du}{dx} = \cos x, \quad \frac{d^2u}{dx^2} = -\sin x, \quad \frac{d^3u}{dx^3} = -\cos x, \text{ \&c.}$$

Making $x = 0$, we obtain for the values of A, A′, &c. in the formula,

$$A = 0, \quad A' = 1, \quad A'' = 0, \quad A''' = -1, \text{ \&c.};$$

thence

$$u = \sin x = \frac{x}{1} - \frac{x^3}{1.2.3} + \frac{x^5}{1.2.3.4.5} - \text{\&c.}$$

2. If

$$u = \cos x,$$

$$\frac{du}{dx} = -\sin x, \quad \frac{d^2u}{dx^2} = -\cos x, \quad \frac{d^3u}{dx^3} = \sin x, \text{ \&c.};$$

in which, making $x = 0$, we obtain

$$A = 1, \quad A' = 0, \quad A'' = -1, \quad A''' = 0, \text{ \&c.};$$

and thence

$$u = \cos x = 1 - \frac{x^2}{1.2} + \frac{x^4}{1.2.3.4} - \text{\&c.}$$

These series, for small values of x, are very converging, and will give with great accuracy the values of $\sin x$ and $\cos x$ for small arcs, and may therefore be used in the calculation of a table of natural sines, &c. Thus, R being unity, we have for the semi-circumference or π, the number 3,14159...; this divided by 180, and the quotient by 60, will give the length of the arc 1′, which value substituted for x in the series, will give the sine and cosine of one minute.

44. We can also develope the arc in terms of its sine, tangent, &c. If

$$x = \sin^{-1} u, \qquad \frac{dx}{du} = \frac{1}{\sqrt{1 - u^2}} \text{......Art. (42)},$$

$$\frac{d^2x}{du^2} = u(1 - u^2)^{-\frac{3}{2}}, \quad \frac{d^3x}{du^3} = (1 - u^2)^{-\frac{3}{2}} + 3u^2(1 - u^2)^{-\frac{5}{2}}, \text{ \&c.}$$

Making $u = 0$, we obtain

$$A = 0 \qquad A' = 1 \qquad A'' = 0 \qquad A''' = 1, \text{ \&c.},$$

and by substitution in Maclaurin's formula

$$x = \sin^{-1} u = u + \frac{u^3}{1.2.3} + \frac{3u^5}{1.2.4.5} + \text{ \&c.}$$

If $u = \frac{1}{2} = \sin 30°$, this series becomes

$$x = \sin^{-1} \frac{1}{2} = 30° = \frac{1}{2} + \frac{1}{1.2.3.2^3} + \frac{3}{1.2.4.5.2^5} + \text{ \&c.},$$

by the summation of which, we find

$$30° = 0{,}52359\text{......},$$

and multiplying by 6, $180° = \pi = 3{,}14159\text{......}$

45. If

$$x = \text{tang}^{-1} u, \qquad \frac{dx}{du} = \frac{1}{1 + u^2} = (1 + u^2)^{-1} \text{......Art. (42)},$$

and the development may be made as in the preceding article; or otherwise thus. Developing $(1 + u^2)^{-1}$ by the binomial formula, we have

$$\frac{dx}{du} = 1 - u^2 + u^4 - u^6 + \&c......(1);$$

and since by differentiation, the exponent of u in each term is diminished by unity, we must have

$$x = \text{A}u + \text{B}u^3 + \text{C}u^5 + \&c.;$$

whence

$$\frac{dx}{du} = \text{A} + 3\text{B}u^2 + 5\text{C}u^4 + \&c......(2).$$

Comparing the coefficients of the like powers of u in (1) and (2),

$$\text{A} = 1,\ 3\text{B} = -1, \text{ and B} = -\frac{1}{3};\ 5\text{C} = 1, \text{and C} = \frac{1}{5}, \&c.;$$

whence

$$x = \text{tang}^{-1}u = u - \frac{u^3}{3} + \frac{u^5}{5} - \frac{u^7}{7} + \&c......(3).$$

If $u = 1 = \text{tang } 45°$, this series becomes

$$x = 45° = 1 - \frac{1}{3} + \frac{1}{5} - \frac{1}{7} + \&c.,$$

which is not sufficiently converging to enable us to determine the length of the arc with accuracy. To obviate this difficulty, we will make use of the principle that the arc 45° is equal to the arc whose tangent is $\frac{1}{2}$, plus the arc whose tangent is $\frac{1}{3}$.*

---

* NOTE.—To prove this principle, take the formula

$$\text{tang}\,(a + b) = \frac{\text{tang } a + \text{tang } b}{1 - \text{tang } a \text{ tang } b}.$$

From equation (3), by the substitution of $\frac{1}{2}$ and $\frac{1}{3}$ for u, we have,

$$\text{tang}^{-1}\frac{1}{2} = \frac{1}{2} - \frac{1}{3.2^3} + \frac{1}{5.2^5} - \frac{1}{7.2^7} + \&c.,$$

$$\text{tang}^{-1}\frac{1}{3} = \frac{1}{3} - \frac{1}{3.3^3} + \frac{1}{5.3^5} - \frac{1}{7.3^7} + \&c.;$$

hence

$$45^\circ = \text{tang}^{-1}\tfrac{1}{2} + \text{tang}^{-1}\tfrac{1}{3} =$$

$$\frac{1}{2} - \frac{1}{3.2^3} + \frac{1}{5.2^5} - \&c. + \frac{1}{3} - \frac{1}{3.3^3} + \frac{1}{5.3^5} - \&c. = 0{,}78539....,$$

which being multiplied by 4 gives $\pi = 3{,}14159......$

DIFFERENTIATION OF FUNCTIONS OF TWO OR MORE VARIABLES.

46. Heretofore our rules for differentiation have been limited to functions of a single variable; it is now proposed to extend them to functions of any number of independent variables.

Let $$u = f(x, y);$$

x and y being entirely independent of each other. The second state of the function will evidently be obtained by giving to both

---

Make tang $a = \frac{1}{2}$ and $a + b = 45^\circ$,

then, tang $45^\circ = 1 = \dfrac{\frac{1}{2} + \text{tang}\, b}{1 - \frac{1}{2}\,\text{tang}\, b}$; whence tang $b = \frac{1}{3}$;

hence $45^\circ = a + b = \text{tang}^{-1}\frac{1}{2} + \text{tang}^{-1}\frac{1}{3}$.

x and y variable increments. First let x receive the increment h; $f(x, y)$ then becomes $f(x + h, y)$, which, (if y for a moment be regarded as constant), may be developed according to the ascending powers of h, by Taylor's formula; whence

$$f(x + h, y) = u + \frac{du}{dx}h + \frac{d^2u}{dx^2}\frac{h^2}{1.2} + \frac{d^3u}{dx^3}\frac{h^3}{1.2.3} + \&c. \ldots\ldots(1),$$

in which $\frac{du}{dx}, \frac{d^2u}{dx^2}$ &c., are the differential coefficients of $u = f(x, y)$ taken under the supposition that x alone is variable, and are evidently all functions of x and y. If in this development we now put $y + k$ for y, we shall obtain in the first member $f(x + h, y + k)$ which is the second state of the function u. The first term of the second member (u), being a function of x and y, will, when for y we put $y + k$, become

$$f(x, y + k) = u + \frac{du}{dy}k + \frac{d^2u}{dy^2}\frac{k^2}{1.2} + \frac{d^3u}{dy^3}\frac{k^3}{1.2.3} + \&c.$$

In the same manner $\frac{du}{dx}$, when for y we put $y + k$, may veloped, and will give, Art. (32),

$$\left(\frac{du}{dx}\right)_{y=y+k} = \frac{du}{dx} + \frac{d\left(\frac{du}{dx}\right)k}{dy} + \frac{d^2\left(\frac{du}{dx}\right)}{dy^2}\frac{k^2}{1.2} + \&c.$$

or reducing

$$\left(\frac{du}{dx}\right)_{y=y+k} = \frac{du}{dx} + \frac{d^2u}{dxdy}k + \frac{d^3u}{dxdy^2}\frac{k^2}{1.2} + \&c.$$

Also

$$\left(\frac{d^2u}{dx^2}\right)_{y=y+k} = \frac{d^2u}{dx^2} + \frac{d^3u}{dx^2dy}k + \frac{d^4u}{dx^2dy^2}\frac{k^2}{1.2} + \&c.$$

$$\left(\frac{d^3u}{dx^3}\right)_{y=y+k} = \frac{d^3u}{dx^3} + \frac{d^4u}{dx^3dy}k + \&c.$$

These values being substituted in the second member of (1) give for the development of the second state of a function of two variables

$$\begin{aligned} f(x+h,\, y+k) = u &+ \frac{du}{dy}k + \frac{d^2u}{dy^2}\frac{k^2}{1.2} + \frac{d^3u}{dy^3}\frac{k^3}{1.2.3} + \&c., \\ &+ \frac{du}{dx}h + \frac{d^2u}{dxdy}hk + \frac{d^3u}{dxdy^2}\frac{hk^2}{1.2} + \&c., \\ &+ \frac{d^2u}{dx^2}\frac{h^2}{1.2} + \frac{d^3u}{dx^2dy}\frac{h^2k}{1.2} + \&c., \\ &+ \frac{d^3u}{dx^3}\frac{h^3}{1.2.3} + \&c. \end{aligned} \quad \ldots\ldots(2).$$

In this development u is the original function; $\frac{du}{dy}$ is its differential coefficient taken under the supposition that y alone varies, and is called, *the partial differential coefficient of the first order taken with respect to y;* $\frac{d^2u}{dy^2}, \frac{d^3u}{dy^3}$ &c. are successive differential coefficients taken under the same supposition, and are called *partial differential coefficients of the second, third, &c. order taken with respect to y.* $\frac{du}{dx}, \frac{d^2u}{dx^2}, \frac{d^3u}{dx^3}$ are obtained from the original function under the supposition that x alone varies, and are called *partial differential coefficients of the first, second, &c., order taken with respect to x;* $\frac{d^2u}{dxdy}$ is obtained by differentiating $\frac{du}{dx}$ with respect to y and dividing the result by dy, and is called a *partial differential co-*

efficient of the second order taken, by differentiating first with respect to x and then with respect to y; and in general $\dfrac{d^{m+n}u}{dx^n dy^m}$ is a partial differential coefficient of the $m + n^{\text{th}}$ order, and is obtained by differentiating first n times with respect to x, and then m times with respect to y.

By an examination of these results we see that, from a function of two variables there are derived two partial differential coefficients of the first order, viz.

$$\frac{du}{dx} \quad \text{and} \quad \frac{du}{dy}$$

three of the second order, viz.

$$\frac{d^2u}{dx^2}, \quad \frac{d^2u}{dxdy}, \quad \frac{d^2u}{dy^2};$$

four of the third order, &c. The expressions

$$\frac{du}{dx}dx, \quad \frac{du}{dy}dy, \quad \frac{d^2u}{dx^2}dx^2, \quad \frac{d^2u}{dxdy}dxdy, \text{ \&c.,}$$

obtained by multiplying the several partial differential coefficients respectively by dx, dy, dx^2, $dxdy$, &c., are called *partial differentials.*

47. If instead of first increasing x by h we increase y by k, we shall obtain

$$f(x, y + k) = u + \frac{du}{dy}k + \frac{d^2u}{dy^2}\frac{k^2}{1.2} + \frac{d^3u}{dy^3}\frac{k^3}{1.2.3} + \text{\&c.,}$$

and if in this we put $x + h$ for x, we shall evidently deduce

$$f(x+h,\,y+k) = u + \frac{du}{dx}h + \frac{d^2u}{dx^2}\frac{h^2}{1.2} + \&c.,$$

$$+ \frac{du}{dy}k + \frac{d^2u}{dydx}kh + \&c.,$$

$$+ \frac{d^2u}{dy^2}\frac{k^2}{1.2} + \&c.,$$

which development must be identical with the one in the preceding article; hence the terms containing the like powers of h and k must be equal to each other, and we must have,

$$\frac{d^2u}{dxdy} = \frac{d^2u}{dydx},\ \frac{d^3u}{dxdy^2} = \frac{d^3u}{dy^2dx},\ldots\ldots\frac{d^{n+m}u}{dx^ndy^m} = \frac{d^{n+m}u}{dy^mdx^n},$$

which shows that we shall obtain the same result whether we differentiate first with reference to x and then with reference to y, or the reverse.

48. Let it now be required to develope the second state of the expression

$$u = x^my^n\ldots\ldots\ldots\ldots(1).$$

Differentiating with reference to x and y, respectively, we obtain

$$\frac{du}{dx} = mx^{m-1}y^n\ldots\ldots(2), \qquad \frac{du}{dy} = nx^my^{n-1}\ldots\ldots(3).$$

Now differentiating (2), first with reference to x, and afterwards with reference to y, we obtain

$$\frac{d^2u}{dx^2} = m(m-1)x^{m-2}y^n\ldots\ldots(4), \qquad \frac{d^2u}{dxdy} = mnx^{m-1}y^{n-1}\ldots\ldots(5).$$

In the same manner by differentiating (3), first with reference to x, and then with reference to y, we obtain

$$\frac{d^2u}{dydx} = mnx^{m-1}y^{n-1} = \frac{d^2u}{dxdy}, \quad \frac{d^2u}{dy^2} = n(n-1)x^my^{n-2} \ldots\ldots(6),$$

and by continuing the differentiation of (4), (5), and (6),

$$\frac{d^3u}{dx^3} = m(m-1)(m-2)x^{m-3}y^n, \frac{d^3u}{dx^2dy} = m(m-1)nx^{m-2}y^{n-1}, \text{\&c.}$$

Substituting these values in the formula of article (46), we have

$$(x+h)^m(y+k)^n = x^my^n + nx^my^{n-1}k + n(n-1)x^my^{n-2}\frac{k^2}{1.2} + \text{\&c.}$$

$$+ mx^{m-1}y^nh + mnx^{m-1}y^{n-1}hk + \text{\&c.}$$

$$+ m(m-1)x^{m-2}y^n\frac{h^2}{1.2} + \text{\&c.}$$

49. Resuming equation (2), Art. (46), and subtracting the primitive state of the function from both members, we obtain

$$f(x+h, y+k) - f(x, y) = \frac{du}{dx}h + \frac{du}{dy}k + \frac{1}{1.2}\left(\frac{d^2u}{dx^2}h^2 + \frac{2d^2u}{dxdy}hk + ..\right).$$

Extending the definition in Art. (8), to functions of two or more variables, we have, after placing for h and k the constants dx and dy, and taking the terms of the first degree with reference to these constants;

$$du = \frac{du}{dx}dx + \frac{du}{dy}dy,$$

that is, the differential of a function of two variables is *equal to the sum of the partial differentials of the function.* It is important to preserve the notation $\frac{du}{dx}dx$ and $\frac{du}{dy}dy$, else the partial differentials might be confounded with the total differential (du).

Examples.

1. If

$$u = ax^2y^3$$

$$\frac{du}{dx}dx = 2axy^3dx, \qquad \frac{du}{dy}dy = 3ax^2y^2dy;$$

hence

$$du = 2axy^3dx + 3ax^2y^2dy.$$

2. If

$$u = \frac{b(a - x^2)^2}{y^2} \qquad du = -\frac{2b}{y^3}(a - x^2)\,[2xydx + (a - x^2)dy].$$

3. If

$$u = \text{tang}^{-1}\frac{x}{y} \qquad du = \frac{ydx - xdy}{y^2 + x^2}.$$

4. Let $u = \frac{ay}{\sqrt{x^2 + y^2}}$. 5. Let $u = x^y$.

50. Having obtained the first differential of a function of two variables, we may from this at once derive the successive differentials. Since

$$du = \frac{du}{dx}dx + \frac{du}{dy}dy,$$

$$d^2u = d\left(\frac{du}{dx}dx\right) + d\left(\frac{du}{dy}dy\right).$$

Differentiating $\frac{du}{dx}dx$, first with reference to x, and then with reference to y, we have,

$$d\left(\frac{du}{dx}dx\right) = \frac{d^2u}{dx^2}dx^2 + \frac{d^2u}{dxdy}dxdy;$$

and in the same way,

$$d\left(\frac{du}{dy}dy\right) = \frac{d^2u}{dydx}dydx + \frac{d^2u}{dy^2}dy^2;$$

whence, since $\frac{d^2u}{dxdy} = \frac{d^2u}{dydx}$......Art. (47),

$$d^2u = \frac{d^2u}{dx^2}dx^2 + 2\frac{d^2u}{dxdy}dxdy + \frac{d^2u}{dy^2}dy^2.$$

Differentiating this result, since

$$d\left(\frac{d^2u}{dx^2}dx^2\right) = \frac{d^3u}{dx^3}dx^3 + \frac{d^3u}{dx^2dy}dx^2dy,$$

$$d\left(\frac{d^2u}{dxdy}dxdy\right) = \frac{d^3u}{dx^2dy}dx^2dy + \frac{d^3u}{dxdy^2}dxdy^2,$$

$$d\left(\frac{d^2u}{dy^2}dy^2\right) = \frac{d^3u}{dy^2dx}dy^2dx + \frac{d^3u}{dy^3}dy^3,$$

we derive

$$d^3u = \frac{d^3u}{dx^3}dx^3 + \frac{3d^3u}{dx^2dy}dx^2dy + \frac{3d^3u}{dxdy^2}dxdy^2 + \frac{d^3u}{dy^3}dy^3.$$

In the same way the differentials of a higher order may be derived.

51. Let us now take the general case in which u is a function of any number of independent variables; that is, let

$$u = f(x, y, z, \&c.)$$

It is plain that we may deduce the development of the second state of this function in precisely the same way as in article (46), by first increasing x and y, then in the result thus obtained increasing z, and in the new result increasing one of the other variables, and so on until each shall have received an increment; we shall thus find

$$f(x+h, y+k, z+l, \&c.) = f(x, y, z, \&c.) + \frac{du}{dx}h + \frac{du}{dy}k + \frac{du}{dz}l + \&c.;$$

whence

$$f(x+h, y+k, z+l, \&c.) - f(x, y, z, \&c.) = \frac{du}{dx}h + \frac{du}{dy}k + \frac{du}{dz}l\ \&c.$$

plus other terms, which will be of the second degree at least, with reference to the increments h, k, l, &c.; we have then as in article (49),

$$du = df(x, y, z, \&c.) = \frac{du}{dx}dx + \frac{du}{dy}dy + \frac{du}{dz}dz + \&c.$$

that is, the differential of a function of any number of variables is equal *to the sum of the partial differentials of the function.*

Example.

If $$u = axy^2z^3,$$

$$du = ay^2z^3dx + 2axyz^3dy + 3axy^2z^2dz.$$

52. If in the development (2), article (46), we make both x and y equal to 0, the first member will become a function of h and k; the first term of the second member, and the different coefficients of h and k, will under the same supposition become constants. Denoting by A what u or $f(x, y)$ becomes when x and y are made 0; by B and B′ what the partial differential coefficients of the first order; by C, C′ and C″ what those of the second order, and by D, D′, D″ and D‴ what those of the third order, become under the same supposition; we obtain

$$f(h, k) = \mathrm{A} + (\mathrm{B}h + \mathrm{B}'k) + \frac{1}{1.2}(\mathrm{C}h^2 + 2\mathrm{C}'hk + \mathrm{C}''k^2)$$

$$+ \frac{1}{1.2.3}(\mathrm{D}h^3 + 3\mathrm{D}'h^2k + \&c.);$$

or since we may change h and k into x and y, we have for the general development of any function of two variables,

$$f(x, y) = \mathrm{A} + (\mathrm{B}x + \mathrm{B}'y) + \frac{1}{1.2}(\mathrm{C}x^2 + 2\mathrm{C}'xy + \mathrm{C}''y^2)$$

$$+ \frac{1}{1.2.3}(\mathrm{D}x^3 + 3\mathrm{D}'x^2y + \&c.).$$

If in development (2), above referred to, we make y and k each equal to 0, u becomes a function of x alone, and we have

$$f(x + h) = u + \frac{du}{dx}h + \frac{d^2u}{dx^2}\frac{h^2}{1.2} + \frac{d^3u}{dx^3}\frac{h^3}{1.2.3} + \&c.$$

which is Taylor's formula.

In the same development, making x, y and k, each equal to 0, and denoting by A, A′, A″, &c. what u, $\frac{du}{dx}$, $\frac{d^2u}{dx^2}$, &c. reduce to under this supposition, we obtain

$$f(h) = A + A'h + A''\frac{h^2}{1.2} + A'''\frac{h^3}{1.2.3} + \&c.,$$

or changing h into x,

$$f(x) = A + A'x + A''\frac{x^2}{1.2} + A'''\frac{x^3}{1.2.3} + \&c.$$

which is Maclaurin's formula.

DIFFERENTIAL EQUATIONS.

53. Every equation containing two variables can, by transposing all the terms into the first member, be represented under the general form

$$f(x, y) = 0 \ldots\ldots\ldots\ldots(1)$$

in which y is an implicit function of x, the latter being usually taken as the independent variable: Or, since by the solution of the equation, the value of y may be found in terms of x, and substituted in (1), this function of x and y may be regarded as a function of x alone, and *may therefore be differentiated as a function of a single variable.* Also, since the relation between y and x is such, that $f(x, y)$ must always be equal to 0, its value is not variable, and can therefore have no difference between any two states. Its differential must then be 0, Art. (14); that is,

$$df(x, y) = 0.$$

To obtain, then, from a given equation *its differential equation, or the equation which expresses the relation between the differentials of the function and variable;* transpose all the terms into one member, differentiate this as a function of a single variable, and place the result equal to 0: Or, if it be not desirable to transpose all the terms into one member, each member, containing either x or

y, or both, may be regarded as a function of x, and the differential of the first will be equal to that of the second, Art. (13).

Since every term of the differential equation thus derived will contain dx or dy, we may divide by dx, and then at once deduce the value of the differential coefficient $\frac{dy}{dx}$.

54. If an equation contain three variables, one will necessarily be a function of the other two, and all the terms being transposed into one member, this member may be regarded as a function of two independent variables, and may be differentiated as in article (49), and the result placed equal to 0. In accordance with the same principles and in precisely the same manner, the differential equation of one containing any number of variables may be derived.

If the differential equation derived by one differentiation be again differentiated, the new differential equation will be of the *second order*, and if this be differentiated we shall have one of the *third order*, and so on.

Examples.

55. 1. If $u = f(x, y) = x^2 + y^2 - R^2 = 0 \ldots\ldots\ldots\ldots(1)$

$$du = 2xdx + 2ydy = 0 \ldots\ldots\ldots\ldots(2),$$

from which, after dividing by dx, we obtain

$$\frac{dy}{dx} = -\frac{x}{y} \ldots\ldots\ldots\ldots(3).$$

Dividing equation (2) by 2, and then differentiating; x, y, and dy, being variable, we have

$$d(xdx) + d(ydy) = 0$$

or $$dx^2 + dy^2 + yd^2y = 0,$$

whence

$$\frac{d^2y}{dx^2} = -\frac{\left(1 + \frac{dy^2}{dx^2}\right)}{y} = -\frac{\left(1 + \frac{x^2}{y^2}\right)}{y} = -\frac{y^2 + x^2}{y^3};$$

since $\frac{dy^2}{dx^2} = \frac{x^2}{y^2}$............equation (3).

Equivalent results may be obtained by differentiating the value $y = \sqrt{R^2 - x^2}$, deduced from equation (1).

2. If $$u = y^2 - 2mxy + x^2 - a^2 = 0 \text{............(1)}$$

$$2ydy - 2mxdy - 2mydx + 2xdx = 0 \text{............(2)};$$

whence

$$\frac{dy}{dx} = \frac{my - x}{y - mx}.$$

Differentiating (2), and dividing by $2dx^2$, we obtain

$$(y - mx)\frac{d^2y}{dx^2} + \frac{dy^2}{dx^2} - 2m\frac{dy}{dx} + 1 = 0;$$

from which after the substitution of the value of $\frac{dy}{dx}$ we may obtain the value of the second differential coefficient.

3. Let $$y^3 - 3axy + x^3 = 0.$$

Equations derived as above, immediately from the primitive equation by differentiation, are named *immediate differential equations.*

56. By the differentiation of equations we may find others which will express the relation between the variables and their differentials, for any values of either or all of the constants. Thus, if we take the equation of the right line

$$y = ax + b \ldots\ldots\ldots\ldots (1),$$

differentiate and divide by dx, we have

$$\frac{dy}{dx} = a \ldots\ldots\ldots\ldots (2),$$

a result which is the same for all values of b. By the substitution of this value of a in equation (1), we have

$$ydx = xdy + bdx,$$

which is the same for all values of a.

Differentiating (2) and dividing by dx, we obtain

$$\frac{d^2y}{dx^2} = 0,$$

which is entirely independent of both a and b.

Take also the equation

$$y^2 = mx + nx^2 \ldots\ldots\ldots\ldots (3).$$

By two differentiations, we get

$$2ydy = mdx + 2nxdx$$

$$dy^2 + yd^2y = ndx^2.$$

By combining the three equations, m and n may readily be eliminated, and an equation obtained which will be entirely independent of them. The result of this elimination is

$$y^2dx^2 + x^2dy^2 + yx^2d^2y - 2yxdydx = 0.$$

Again, by differentiating the equation

$$y^3 - 2ax^2 + a^2 = 0,$$

and eliminating a, we obtain

$$16yx^2dx^2 - 24x^3dydx + 9y^2dy^2 = 0.$$

And in general, all the constants of any equation may be eliminated by differentiating it as many times as there are constants. The differential equations thus obtained, with the given equation, make one more than the number of constants to be eliminated; an equation may therefore be derived which will be freed from these constants. Equations thus obtained are properly *the differential equations of the species of lines*, one of which is represented by the given equation, since being independent of the constants they are evidently the same for all lines of the same kind referred to the same co-ordinate axes.

57. By differentiation we may free an equation of exponents, as in the example

$$u = v^n,$$

$$du = nv^{n-1}dv, \qquad \text{or} \qquad vdu = nv^n dv,$$

and finally

$$vdu = nudv.$$

Or thus,

$$lu = lv^n; \qquad \text{whence} \qquad lu = nlv,$$

$$\frac{du}{u} = \frac{ndv}{v}, \qquad \text{or} \qquad vdu = nudv.$$

58. The Differential Calculus enables us also to eliminate, from an equation containing three variables, an arbitrary function of either two, the form of which may be entirely unknown. Thus if

$$u = \mathrm{F}\,(f\,[x, y]),$$

the form of the function designated by the symbol F being arbitrary, we can find a differential equation expressing a relation between x, y and the partial differential coefficients $\frac{du}{dx}, \frac{du}{dy}$, which will be the same, no matter what the form of the function F may be.

Make $$f(x, y) = z \ldots\ldots\ldots\ldots(1),$$

then

$$u = \mathrm{F}(z), \qquad du = \mathrm{F}'(z)dz \ldots\ldots(2).$$

Differentiating (1), first with reference to x and then with reference to y, and substituting the values of dz thus obtained in (2), we get

$$\frac{du}{dx} = \mathrm{F}'(z)\,\frac{dz}{dx} \ldots\ldots(3), \qquad \frac{du}{dy} = \mathrm{F}'(z)\,\frac{dz}{dy} \ldots\ldots(4),$$

from which $\mathrm{F}'(z)$ may be eliminated, and the resulting equation, between x, y, $\frac{du}{dx}$ and $\frac{du}{dy}$, will be the differential equation required. Such equations are called *partial differential equations.*

To illustrate, suppose

1. $$f(x, y) = ax + by \quad \text{and} \quad u = \mathrm{F}(ax + by).$$

Place $ax + by = z$, then $\frac{dz}{dx} = a$, and $\frac{dz}{dy} = b$.

These values in equations (3) and (4), give

$$\frac{du}{dx} = F'(z)a \qquad \frac{du}{dy} = F'(z)b;$$

whence

$$\frac{\frac{du}{dx}}{a} = \frac{\frac{du}{dy}}{b}$$

and finally

$$a\frac{du}{dy} - b\frac{du}{dx} = 0.$$

2. Let

$$f(x, y) = x^2 + y^2 = z \quad \text{and} \quad u = F(x^2 + y^2).$$

Differentiating z, we find

$$\frac{dz}{dx} = 2x \quad \text{and} \quad \frac{dz}{dy} = 2y;$$

whence

$$\frac{du}{dx} = F'(z)2x \quad \text{and} \quad \frac{du}{dy} = F'(z)2y.$$

from which, by eliminating $F'(z)$,

$$x\frac{du}{dy} - y\frac{du}{dx} = 0.$$

3. Let $$f(x, y) = \frac{x}{y}.$$

VANISHING FRACTIONS.

59. In the discussion of the results obtained by the application of the Calculus, we often meet with expressions which, for a particular value of the variable, become $\frac{0}{0}$. This, although in general the algebraic symbol of an indeterminate quantity, does not indicate such a quantity in the particular cases referred to. As in the example,

$$\frac{ax - x^2}{a^2 - x^2}$$

which becomes $\frac{0}{0}$ when $x = a$; if we divide both numerator and denominator by the common factor $a - x$, we obtain

$$\frac{x}{a + x}$$

and this, when $x = a$, reduces to $\frac{1}{2}$, which is the true value of the fraction in the particular case.

Expressions of this kind are called *vanishing fractions*, and reduce to $\frac{0}{0}$ in consequence of the existence of a factor common to both terms; which factor becomes 0 under the particular supposition.

All such fractions may be represented generally by the expression

$$\frac{\mathrm{P}(x - a)^m}{\mathrm{Q}(x - a)^n};$$

in which P and Q are functions of x.

There are three cases:

1. When $m = n$, the fraction becomes

$$\frac{P(x-a)^m}{Q(x-a)^m} = \frac{P}{Q}.$$

2. When $m > n$, it may be put under the form

$$\frac{P(x-a)^{m-n}}{Q},$$

$m - n$ being positive; and this, when $x = a$, becomes

$$\frac{0}{Q_{x=a}} = 0.$$

3. When $m < n$, the fraction may be put under the form

$$\frac{P}{Q(x-a)^{n-m}};$$

$n - m$ being positive, and this, when $x = a$, becomes

$$\frac{P_{x=a}}{0} = \infty.$$

60. Whenever the common factor is evident, the simplest method of obtaining the true value of the fraction is to strike it out, and then put for the variable its particular value. But as in most cases it is not easy to detect this factor, other methods become necessary.

Let $\frac{r}{s}$ be a vanishing fraction, r and s being functions of x, and let a be the particular value which substituted for x reduces the fraction to $\frac{0}{0}$.

It is plain that, if we substitute $a + h$ for x, and after reduction make $h = 0$, it will amount only to the substitution of a for x. Suppose this substitution made, and that in the result both numerator and denominator are arranged so that the exponents of h shall increase from left to right, we then have

$$\left(\frac{r}{s}\right)_{x=a+h} = \frac{Ah^m + Bh^{m'} + \&c.}{A'h^n + B'h^{n'} + \&c.}$$

in which A, A′, B, B′, m, n, &c. are constants. After reducing this fraction to its lowest terms, by dividing both numerator and denominator by that power of h which is indicated by the smallest exponent, we shall have one of three cases.

1. If $m = n$

$$\left(\frac{r}{s}\right)_{x=a+h} = \frac{A + Bh^{m'-m} + \&c.}{A' + B'h^{n'-n} + \&c.}$$

2. If $m > n$

$$\left(\frac{r}{s}\right)_{x=a+h} = \frac{Ah^{m-n} + \&c.}{A' + \&c.}$$

3. If $m < n$

$$\left(\frac{r}{s}\right)_{x=a+h} = \frac{A + \&c.}{A'h^{n-m} + \&c.}$$

Now making $h = 0$, we have for the true value in the three cases,

1. $\left(\frac{r}{s}\right)_{x=a} = \frac{A}{A'}$. 2. $\left(\frac{r}{s}\right)_{x=a} = \frac{0}{A'} = 0$.

3. $\left(\frac{r}{s}\right)_{x=a} = \frac{A}{0} = \infty$.

Whence we derive the general rule. *For the variable, substitute that value which causes the fraction to reduce to $\frac{0}{0}$, plus an increment; reduce the result to its simplest form, and then make the* *increment equal to* 0. The final result will be the true value of the fraction for the particular value of the variable, and may be finite, zero, or infinite.

Examples.

1. Take the fraction

$$\frac{(x^2 - a^2)^{\frac{3}{2}}}{(x - a)^{\frac{3}{2}}},$$

which becomes $\frac{0}{0}$ when $x = a$.

For x, put $a + h$, the primitive fraction then becomes

$$\frac{(2ah + h^2)^{\frac{3}{2}}}{h^{\frac{3}{2}}}$$

Dividing both terms by $h^{\frac{3}{2}}$, we obtain

$$(2a + h)^{\frac{3}{2}},$$

which, when $h = 0$, becomes $(2a)^{\frac{3}{2}}$, the true value.

In this case the common factor $(x - a)^{\frac{3}{2}}$ is evident; striking it out, we have

$$(x + a)^{\frac{3}{2}}$$

which becomes $(2a)^{\frac{3}{2}}$, when $x = a$.

2. Take the fraction

$$\frac{m \sin^{-1} \frac{x}{a}}{x},$$

which becomes $\frac{0}{0}$ when $x = 0$.

For x, put $0 + h$, or h, we then obtain

$$\frac{m \sin^{-1} \dfrac{h}{a}}{h} = m \frac{\left(\dfrac{h}{a} + \dfrac{h^3}{1.2.3a^3} + \&c.\right)}{h} \text{......Art. (44)},$$

or

$$\frac{m \sin^{-1} \dfrac{h}{a}}{h} = m\left(\frac{1}{a} + \frac{h^2}{1.2.3a^3} + \&c.,\right)$$

which, when $h = 0$, gives

$$\left(\frac{m \sin^{-1} \dfrac{x}{a}}{x}\right)_{x=0} = \frac{m}{a}.$$

The common factor in this case is x, as may be shown by developing $m \sin^{-1} \dfrac{x}{a}$, as in article (44).

61. Another rule may be thus deduced.

If the vanishing fraction, as in the preceding article, be

$$u = \frac{r}{s}; \qquad \text{then} \qquad r = us$$

$$dr = uds + sdu;$$

in which, if we make $x = a$, we shall have (since $s_{x=a} = 0$),

$$(dr)_{x=a} = (uds)_{x=a};$$

whence

$$u_{x=a} = \left(\frac{r}{s}\right)_{x=a} = \frac{(dr)_{x=a}}{(ds)_{x=a}} \dots\dots\dots\dots (1),$$

for the true value of the fraction in the particular case.

If $(dr)_{x=a} = 0$ this value is 0.

If $(ds)_{x=a} = 0$ it is ∞.

If both are 0 at the same time, the second member of (1) becomes $\frac{0}{0}$, and $\frac{dr}{ds}$ is a new vanishing fraction; then, as above, we take the differentials of both its terms, put a for x, and thus obtain

$$u_{x=a} = \frac{(d^2r)_{x=a}}{(d^2s)_{x=a}}.$$

If this again becomes $\frac{0}{0}$, we continue the same process, and have

$$u_{x=a} = \frac{(d^3r)_{x=a}}{(d^3s)_{x=a}},$$

and so on. The rule may then be thus enunciated. *Take the differentials of the numerator and denominator; in each, substitute that value of the variable which reduces the original fraction to $\frac{0}{0}$; if both do not reduce to 0 or infinity; what the former becomes divided by what the latter becomes, will be the true value of the fraction. If both reduce to 0, take the second differentials, and* *make the same substitution ; or continue the differentiation, &c. until two differentials of the same order are obtained, both of which do not become* 0 *or infinity ; what one becomes divided by what the other becomes, will be the true value of the fraction.*

It should be observed, that the effect of the application of this rule is, at each differentiation, to diminish by unity the exponent of the factor which causes the fraction to reduce to $\frac{0}{0}$, Art. (27). If the exponents of this factor in the numerator and denominator are fractional, and not contained between the same two consecutive whole numbers, it is plain that the least one will be reduced to a negative number by a less number of differentiations than will be required by the other. The differential of that term of the fraction which contains it, will then, by the substitution of the particular value of the variable, reduce to infinity, while that of the other reduces to 0, and the true value of the fraction will be either

$$\frac{\infty}{0} = \infty \qquad \text{or} \qquad \frac{0}{\infty} = 0.$$

If however, these exponents are contained between the same two consecutive whole numbers, they will become negative by the same number of differentiations, and the differentials of both terms of the fraction, reduce to infinity at the same time ; as will the successive differentials. In this *the only failing case* of the rule, we shall not be able, by its application, to obtain the true value of the fraction, but must fall back upon the general rule, Art. (60). As an illustration of this, we may refer to example 1, article (60), in which the second differentials, and all which follow, become infinite when $x = a$.

Examples

If

$$\frac{r}{s} = \frac{x^n - 1}{x - 1},$$

which becomes $\frac{0}{0}$ when $x = 1$

$$dr = nx^{n-1}dx, \qquad ds = dx,$$

$$\left(\frac{r}{s}\right)_{x=1} = \frac{(dr)_{x=1}}{(ds)_{x=1}} = \left(\frac{nx^{n-1}}{1}\right)_{x=1} = n.$$

2. If

$$\frac{r}{s} = \frac{1 - \sin x}{\cos x}$$

which becomes $\frac{0}{0}$ when $x = \frac{\pi}{2}$,

$$dr = -\cos x dx, \qquad ds = -\sin x dx,$$

$$\left(\frac{r}{s}\right)_{x=\frac{\pi}{2}} = \left(\frac{\cos x}{\sin x}\right)_{x=\frac{\pi}{2}} = 0.$$

3. If $$\frac{r}{s} = \frac{ax^2 - 2acx + ac^2}{bx^2 - 2bcx + bc^2},$$

$$dr = (2ax - 2ac)dx, \qquad ds = (2bx - 2bc)dx,$$

both of which reduce to 0, when $x = c$. Differentiating again,

$$d^2r = 2adx^2, \qquad d^2s = 2bdx^2,$$

and

$$\left(\frac{r}{s}\right)_{x=c} = \frac{a}{b}.$$

4. Take $\dfrac{a^x - b^x}{x}$ when $x = 0$. Ans. $la - lb$.

5. $\dfrac{ml\left(1 + \dfrac{x}{a}\right)}{x}$ $x = 0$. " $\dfrac{m}{a}$.

6. $\dfrac{1 - \sin x + \cos x}{\sin x + \cos x - 1}$ $x = \dfrac{\pi}{2}$.

7. $\dfrac{a - x - ala + alx}{a - \sqrt{2ax - x^2}}$ $x = a$.

8. $\dfrac{x^x - x}{1 - x + lx}$ $x = 1$.

9. $\dfrac{x - 2 \sin x}{x \sin x}$ $x = 0$.

62. We sometimes meet with the product of two factors, one of which becomes 0, and the other ∞, for a particular value of the variable. Let rt be such a product, in which r becomes 0, and t infinite. It may be written

$$rt = \frac{r}{\frac{1}{t}},$$

which, for the particular value, becomes $\frac{0}{0}$. Its value may then be determined as in the preceding articles.

Example.

Let $rt = (1 - x) \operatorname{tang} \dfrac{\pi x}{2}$, when $x = 1$.

Writing it under the proposed form, we have

$$rt = \frac{1 - x}{\dfrac{1}{\text{tang } \dfrac{\pi x}{2}}} = \frac{1 - x}{\cot \dfrac{\pi x}{2}},$$

the true value of which, when $x = 1$, is $\frac{2}{\pi}$.

63. The fraction $\frac{r}{s}$ may become $\frac{\infty}{\infty}$, in which case it may be written

$$\frac{r}{s} = r \times \frac{1}{s},$$

which becomes $\infty \times \frac{1}{\infty} = \infty \times 0$, and may then be treated as in the preceding article.

64. Sometimes also, we find expressions which become $\infty - \infty$.

Let $$\frac{1}{r} - \frac{1}{s},$$

be such an expression, r and s becoming 0. It may be written

$$\frac{1}{r} - \frac{1}{s} = \frac{s - r}{rs},$$

which will reduce to $\frac{0}{0}$. For an example, take

$$\frac{x}{\cot x} - \frac{\pi}{2 \cos x},$$

which becomes $\infty - \infty$, when $x = \frac{\pi}{2}$. By reduction we obtain

$$\frac{x \sin x - \frac{\pi}{2}}{\cos x};$$

the true value of which is, -1, when $x = \frac{\pi}{2}$.

MAXIMA AND MINIMA.

65. A function is at a maximum state, or *a maximum, when it is greater than the state which immediately precedes, and greater also than the state which immediately follows it;* and *a minimum, when it is less than both of these states.*

Thus, if u be a function of x, and x be decreased so as to give the next preceding state to u, denoted by u'', and then increased, by the same quantity, so as to give the next succeeding state u'; if u be greater than both u'' and u' it will be *a maximum;* if less, *a minimum.*

66. If u is a function of x, and x supposed to be increasing, it is evident that when passing from the preceding states to its maximum, u must *increase* as x increases, that is, be an *increasing function* of x; and when passing from its maximum to the succeeding states, it must *decrease* as x increases, that is, be *a decreasing function* of x. In the first case, Art. (12), the sign of its first differential coefficient must be positive, and in the second, negative; therefore at the maximum state *the first differential coefficient must change its sign from plus to minus.* For a similar reason at a minimum state, the first differential coefficient must

change its sign *from minus to plus.* But as a quantity can change its sign only by becoming zero or infinity, it follows that no value of the variable will give a maximum or minimum value to the function, unless the same value reduces the first differential coefficient to zero or infinity.

The roots of the two equations

$$\frac{du}{dx} = 0 \ldots\ldots (1), \qquad \text{and} \qquad \frac{du}{dx} = \infty \text{ or } \frac{dx}{du} = 0 \ldots\ldots (2),$$

will then give all the values of x, which can possibly make u a maximum or a minimum. After having obtained these roots, let each, first with an infinitely small decrement, and then with an infinitely small increment, be substituted in the given function; if both the results are less than the one obtained by substituting the root, the latter will be a maximum; if both are greater, a minimum.

Or if it be more convenient, let each of these roots, with an infinitely small decrement and increment, be successively substituted in the first differential coefficient; if the first result be positive, and the second negative, the root will make the function a maximum; if the reverse, a minimum. If the two results have the same sign, the root under consideration will give neither a maximum nor a minimum.

Since equations (1) and (2) may give several roots which will fulfil the required conditions, there may be more than one maximum or minimum state of the same function.

Examples.

1. If $$u = a + (x - b)^2 \ldots\ldots\ldots\ldots (3),$$

$$\frac{du}{dx} = 2(x - b) \qquad \text{and} \qquad \frac{dx}{du} = \frac{1}{2(x - b)}.$$

Placing $\frac{du}{dx} = 0$, we have

$$2(x - b) = 0; \qquad \text{whence} \qquad x = b$$

If in equation (3) we substitute first, $b - h$ for x, and then $b + h$, we have

$$u'' = a + h^2 \qquad \text{and} \qquad u' = a + h^2$$

both of which for all values of h are *greater* than $u = a$, the result obtained by substituting b for x; *hence* $u = a$ *is a minimum.*

The only value of x which will reduce $\frac{dx}{du}$ to 0 is $x = \infty$; there is then no finite value of x which will satisfy this condition, hence $x = b$ gives the only minimum state, and there is no maximum.

2. If

$$u = a - (x - b)^{\frac{2}{3}} \ldots\ldots\ldots\ldots (4)$$

$$\frac{du}{dx} = \frac{-2}{3(x - b)^{\frac{1}{3}}}, \qquad \text{and} \qquad \frac{dx}{du} = \frac{-3(x - b)^{\frac{1}{3}}}{2}.$$

Placing $\frac{du}{dx} = 0$ we obtain $x = \infty$, which gives no finite solution.

Placing $\frac{dx}{du} = 0$, we have

$$3(x - b) = 0; \qquad \text{whence} \qquad x = b.$$

If then in (4), we substitute first $b - h$, and then $b + h$, for x, we have

$$u'' = a - h^{\frac{2}{3}} \qquad \text{and} \qquad u' = a - h^{\frac{2}{3}},$$

both of which are less than $u = a$, the result of the substitution of b for x; $u = a$ *is then a maximum and the only one, and there is no minimum.*

If in the first differential coefficients in the above examples we substitute $b - h$ and $b + h$ for x, we obtain in the first, for $b - h$ a negative, and for $b + h$ a positive result, and in the second the reverse, as it should be.

67. When the states which immediately precede and follow the maximum or minimum state of u, can be deduced from Taylor's formula, a more convenient rule may be applied. To demonstrate it; let

$$u = f(x),$$

then

$$u' = f(x + h) \qquad u'' = f(x - h),$$

and by Taylor's formula

$$\left.\begin{aligned} u' - u &= \frac{du}{dx} h + \frac{d^2u}{dx^2} \frac{h^2}{1.2} + \frac{d^3u}{dx^3} \frac{h^3}{1.2.3} + \&c. \\ u'' - u &= -\frac{du}{dx} h + \frac{d^2u}{dx^2} \frac{h^2}{1.2} - \frac{d^3u}{dx^3} \frac{h^3}{1.2.3} + \&c. \end{aligned}\right\} \text{Art. (34.)}$$

In order that u be a maximum it must be greater than both u' and u'', that is, the second members of the above equations, for an infinitely small value of h, must be negative; and for a minimum the reverse. But for any value of h less than the one referred to in article (11), (and of course when h is infinitely small), the signs of the series will be the same as the signs of their first terms; but these terms have contrary signs, hence there can be neither maximum nor minimum unless the first term of each series be 0, which requires that $\frac{du}{dx} = 0$. The roots of this equation will then, in the case under consideration, give all the values of x which can possibly make u either a maximum or minimum.

Let a be one of these roots, and let it be substituted for x in the two series, then, since $\left(\frac{du}{dx}\right)_{x=a} = 0$, we have

$$(u' - u)_{x=a} = \left(\frac{d^2u}{dx^2}\right)_{x=a} \frac{h^2}{1.2} + \left(\frac{d^3u}{dx^3}\right)_{x=a} \frac{h^3}{1.2.3} + \&c.$$

$$(u'' - u)_{x=a} = \left(\frac{d^2u}{dx^2}\right)_{x=a} \frac{h^2}{1.2} - \left(\frac{d^3u}{dx^3}\right)_{x=a} \frac{h^3}{1.2.3} + \&c.$$

The signs of the series now depend upon that of $\left(\frac{d^2u}{dx^2}\right)_{x=a}$, and will both be negative, and $u_{x=a}$ a maximum, if $\left(\frac{d^2u}{dx^2}\right)_{x=a}$ *is negative;* and the reverse if this is *positive.* But if $\left(\frac{d^2u}{dx^2}\right)_{x=a} = 0$, the signs of the series will again be contrary, and there can be neither maximum nor minimum unless $\left(\frac{d^3u}{dx^3}\right)_{x=a} = 0$, in which case the signs will be the same as that of $\left(\frac{d^4u}{dx^4}\right)_{x=a}$: And in general, if there be either a maximum or minimum, the first differential coefficient which does not reduce to 0 when $x = a$, must be of an even order, *negative for a maximum*, and *positive for a minimum.* Whence to determine the maximum or minimum states of a given function. *Find its first differential coefficient and place it equal to* 0 ; *substitute each of the real roots of the equation thus formed, in the second differential coefficient. Each one which gives a negative result, will when substituted in the function make it a maximum, and each which gives a positive result will make it a minimum. If either reduce the second differential coefficient to* 0, *substitute in the third, fourth, &c. until one be obtained which does not reduce to* 0. *If this be of an odd order, the root will correspond to neither a maximum nor minimum; if of an even order and negative, there will be a corresponding maximum; if positive, a minimum.*

Examples.

1. If
$$u = \frac{x^3}{3} + ax^2 - 3a^2x,$$

$$\frac{du}{dx} = x^2 + 2ax - 3a^2, \qquad \frac{d^2u}{dx^2} = 2x + 2a \ldots\ldots\ldots\ldots (1).$$

Placing the value of $\frac{du}{dx} = 0$, we have

$$x^2 + 2ax - 3a^2 = 0,$$

the roots of which are $x = a$, and $x = -3a$. The first substituted in (1) gives $4a$, which being positive, indicates a minimum. The second substituted in (1) gives $-4a$, which indicates a maximum. Substituting the roots in the given function, we have for the minimum $u = -\frac{5a^3}{3}$, and for the maximum $u = 9a^3$.

2. If
$$u = 2x^4 + a^3x,$$

$$\frac{du}{dx} = 8x^3 + a^3, \qquad \frac{d^2u}{dx^2} = 24x^2 \ldots\ldots\ldots\ldots (2).$$

Placing the value of $\frac{du}{dx} = 0$, we have

$$8x^3 + a^3 = 0; \qquad \text{whence} \qquad x = -\frac{a}{2}.$$

This value of x in (2) gives $6a^2$, and indicates a minimum, which is $u = -\frac{3a^4}{8}$.

68. Let $$v = Au,$$

u being any function of x. By differentiation, &c., we have

$$\frac{dv}{dx} = A\frac{du}{dx}, \qquad \frac{d^2v}{dx^2} = A\frac{d^2u}{dx^2};$$

from which it appears, that those values of x, which make $\frac{du}{dx} = 0$, will also make $\frac{dv}{dx} = 0$, and the reverse. Also, that any of these values, when substituted in the second differential coefficients, will give results affected with the same sign. Hence every value of x which will make u a maximum or minimum will make Au a maximum or minimum. *Therefore a constant positive factor may be omitted during the search for those values of the variable corresponding to a maximum or minimum.*

To illustrate, take the example

$$\frac{b}{a}(2ax - x^2)\ldots\ldots(1).$$

Omitting the constant factor, we may write

$$u = 2ax - x^2,$$

$$\frac{du}{dx} = 2a - 2x, \qquad \frac{d^2u}{dx^2} = -2.$$

Placing $\frac{du}{dx} = 0$, we find $x = a$, which in (1) gives the maximum value ab.

69. Let $$v = u^n,$$

u and v being functions of x, and n entire. Then

$$\frac{dv}{dx} = nu^{n-1}\frac{du}{dx},$$

$$\frac{d^2v}{dx^2} = nu^{n-1}\frac{d^2u}{dx^2} + n(n-1)u^{n-2}\frac{du^2}{dx^2}.$$

Now every value of x which will make $\frac{du}{dx} = 0$, will also make $\frac{dv}{dx} = 0$; and if the same value makes nu^{n-1} *positive*, it will give to $\frac{d^2v}{dx^2}$ the same sign as $\frac{d^2u}{dx^2}$ (since $\frac{du^2}{dx^2} = 0$); that is, if it makes u a maximum or minimum it will make v a maximum or minimum. If it makes nu^{n-1} *negative*, it will give to $\frac{d^2v}{dx^2}$ a sign contrary to that of $\frac{d^2u}{dx^2}$; that is, if it makes u a maximum, it will make v a minimum, and the reverse.

All values of x, however, which will make $v = u^n$ a maximum or minimum, will not necessarily make u a maximum or minimum, for the equation

$$\frac{dv}{dx} = nu^{n-1}\frac{du}{dx} = 0,$$

may be satisfied by making either

$$nu^{n-1} = 0, \qquad \text{or} \qquad \frac{du}{dx} = 0.$$

Those values of x which satisfy the first, and not the second of these equations, will make u neither a maximum nor minimum, but may make $v = u^n$ a maximum or minimum. As in the example,

$$v = (a^3 - x^3)^2 = u^2,$$

$$dv = 2udu \qquad \frac{dv}{dx} = 2u\frac{du}{dx}.$$

We may make $\frac{dv}{dx} = 0$, by placing either

$$2u = 2(a^3 - x^3) = 0; \qquad \text{whence} \qquad x = a,$$

or

$$\frac{du}{dx} = -3x^2 = 0, \qquad \text{"} \qquad x = 0.$$

The value $x = a$ evidently makes v a minimum, but as it does not reduce $\frac{du}{dx} = -3x^2$ to 0, it will make u neither a maximum nor minimum.

The value $x = 0$ answers to neither a maximum nor a minimum. As the corresponding power of a radical expression is formed by omitting the radical sign, we may, in accordance with the above principles, omit it, and seek those values of the variable which will make the power a maximum or minimum. We are sure thus to get all the values which will make the root a maximum or minimum. Care should be taken, however, not to use any of those which belong only to the power.

70. In a manner similar to the above, it may be shown that any value of the variable which will render u a maximum or minimum will also render $\log u$ and a^u a maximum or minimum.

71. It often happens that the first differential coefficient is composed of two or more variable factors, each of which, when placed equal to 0, may give values of the variable, corresponding to maximum or minimum states of the function. Let

$$\frac{du}{dx} = \text{XX}',$$

be such a coefficient, X being 0 when $x = a$. Then

$$\frac{d^2u}{dx^2} = \text{X}\frac{d\text{X}'}{dx} + \text{X}'\frac{d\text{X}}{dx};$$

or since $\text{X} = 0$ when $x = a$,

$$\left(\frac{d^2u}{dx^2}\right)_{x=a} = \left(\text{X}'\frac{d\text{X}}{dx}\right)_{x=a}.$$

That is, to obtain the corresponding value of the second differential coefficient; *multiply the differential coefficient of that factor which is* 0, *by the other factors, and then substitute the particular value of the variable. To illustrate, let*

$$u = x^2(x - a)^6,$$

$$\frac{du}{dx} = 2x(x - a)^5(4x - a),$$

which is equal to 0, when

$$2x = 0; \qquad \text{whence} \qquad x = 0 \ldots\ldots(1).$$

$$(x - a)^5 = 0; \qquad \text{"} \qquad x = a \ldots\ldots(2).$$

$$(4x - a) = 0; \qquad \text{"} \qquad x = \frac{a}{4} \ldots\ldots(3).$$

Taking the first factor $2x$, and multiplying its differential coefficient by the other factors, we obtain the expression

$$2(x-a)^5(4x-a);$$

from which, by making $x = 0$, we obtain

$$\left(\frac{d^2u}{dx^2}\right)_{x=0} = 2a^6;$$

which indicates a minimum.

Multiplying the differential coefficient of the third factor $4x - a$, by the others, and making $x = \frac{a}{4}$, we obtain a negative result, which indicates a maximum.

The second value of x reduces $\frac{d^2u}{dx^2}$ to 0, but will make $\frac{d^6u}{dx^6}$ positive, and give a minimum, Art. (67).

72. If the function be implicit, we have only to find its differential coefficient as in article (17) or (53), and proceed as with an explicit function. To illustrate, take the example

$$y^2 - 2mxy + x^2 - a^2 = 0 \ldots\ldots(1),$$

and let it be required to find the value of x which will make y a maximum or minimum. By differentiating as in article (53), we obtain

$$2ydy - 2mxdy - 2mydx + 2xdx = 0;$$

whence

$$\frac{dy}{dx} = \frac{my - x}{y - mx} \ldots\ldots(2).$$

Placing this equal to 0, we have

$$my - x = 0; \qquad \text{whence} \qquad x = my,$$

which, in equation (1), gives

$$y = \frac{a}{\sqrt{1 - m^2}}; \qquad \text{whence} \qquad x = \frac{ma}{\sqrt{1 - m^2}}.$$

Differentiating the factor $my - x$, equation (2), dividing by dx, and multiplying by $\frac{1}{y - mx}$, Art. (71), we obtain the expression

$$\frac{1}{y - mx}\left(m\frac{dy}{dx} - 1\right),$$

which, by the substitution of the values of y and x, (since then $\frac{dy}{dx} = 0$), becomes

$$\frac{-1}{a\sqrt{1 - m^2}},$$

and indicates a maximum.

73. The only difficulty in the application of the preceding principles to the solution of problems, consists in obtaining a convenient algebraic expression for the function whose maximum or minimum state is required. No general rule can well be given by which this expression can be found. In order to indicate as clearly as possible the methods to be pursued, we will give the solution of several cases differing from each other.

1. Required the dimensions of the maximum cylinder, which can be inscribed in a given right cone.

Suppose a cylinder inscribed, as represented in the figure. Let

$$VA = a, \quad BA = b, \quad VC = x, \quad CO = y$$

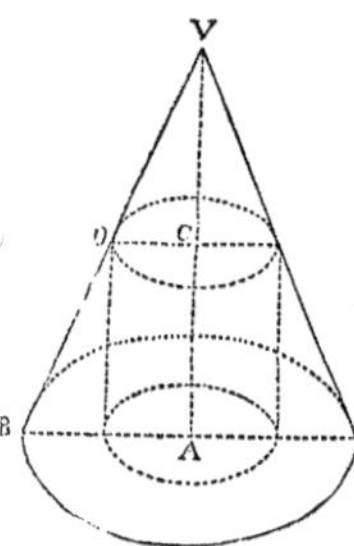

then $AC = a - x$, and the solidity of the cylinder, which we denote by v, is equal to

$$\pi y^2(a - x) \ldots\ldots\ldots\ldots (1).$$

From the similar triangles VCO and VAB, we have the proportion

$$x : y :: a : b; \qquad \text{whence} \qquad y = \frac{bx}{a}.$$

Substituting this value in (1), we have

$$v = \frac{\pi b^2}{a^2} x^2(a - x) \ldots\ldots\ldots\ldots (2).$$

Omitting the constant factor, Art. (68), we may write

$$u = ax^2 - x^3;$$

whence

$$\frac{du}{dx} = 2ax - 3x^2, \qquad \frac{d^2u}{dx^2} = 2a - 6x \ldots\ldots (3).$$

Placing $\dfrac{du}{dx} = 0$, we find the roots $x = 0$, and $x = \frac{2}{3}a$. The second value of x in (3) gives $-2a$, and therefore will make v a maximum, which is $\dfrac{4\pi ab^2}{27}$.

For the altitude of the maximum cylinder, we have $a - x = \frac{1}{3}a$, and for the radius of the base $y = \frac{2}{3}b$.

The first value of x in (3) gives $2a$, which indicates a minimum, which is evidently $v = 0$.

2. Required to draw a tangent to the given quadrant ABD, so that the triangle CFG shall be a minimum.

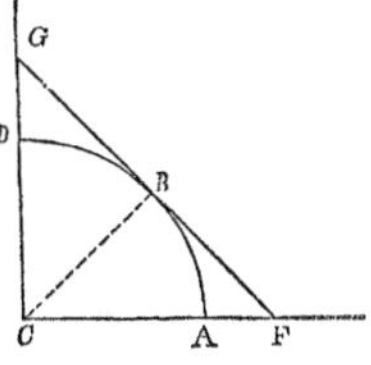

Let CB = R, FB = x, BG = y; then FG = $x + y$. The area of the triangle is equal to $\frac{1}{2}$CB × FG, which since $\frac{1}{2}$CB is constant, will be a minimum when FG is a minimum, Art. (68). In the right angled triangle CFG, since CB is perpendicular to FG, we have

$$R^2 = xy; \qquad \text{whence} \qquad y = \frac{R^2}{x},$$

and

$$FG = u = x + \frac{R^2}{x}$$

$$\frac{du}{dx} = 1 - \frac{R^2}{x^2} = \frac{x^2 - R^2}{x^2},$$

which, being placed equal to 0, gives $x = R$, and $y = R$.

Hence the angle BCF = 45°. Obtaining the corresponding value of $\frac{d^2u}{dx^2}$, as in Art. (71), we find for a result $\frac{2}{R}$.

3. The whole surface of a right cylinder being given, it is required to find the radius of the base, and altitude, when the solidity is a maximum.

Let m^2 = the surface, x = the radius of the base, and z = the altitude, then

$$v = \pi x^2 z.$$

But

$$m^2 = 2\pi xz + 2\pi x^2; \qquad \text{whence} \qquad z = \frac{m^2 - 2\pi x^2}{2\pi x};$$

therefore

$$v = \frac{m^2 x}{2} - \pi x^3,$$

and $x = \sqrt{\frac{m^2}{6\pi}}$, and $z = 2\sqrt{\frac{m^2}{6\pi}}$, when v is a maximum.

4. Required to divide a given quantity a, into two parts, such that the mth power of one, multiplied by the nth power of the other, shall be a maximum.

If x = one of the parts, then $x = \frac{ma}{m+n}$.

5. In a given triangle, it is required to inscribe a maximum rectangle.

The altitude of the rectangle = ½ altitude of triangle.

6. A certain quantity of water being given, it is required to find the relation between the radius of the base and altitude of a cylindrical vessel, open at the top, which shall just hold the water and have its interior surface a minimum.

The radius = the altitude.

7. Required the maximum rectangle which can be inscribed in a circle.

Each side = $R\sqrt{2}$.

8. Required the maximum cone which can be inscribed in a given sphere.

9. Required the minimum triangle that can be circumscribed about a given portion of a semi-parabola.

10. Required the maximum cylinder that can be inscribed in a given ellipsoid of revolution.

11. Required the axis of the maximum parabola that can be cut from a given right cone.

12. Required the minimum value of y in the equation $y = x^x$.

MAXIMA AND MINIMA OF FUNCTIONS OF TWO OR MORE VARIABLES.

74. A function of two or more variables is a maximum when it is greater, and a minimum when it is less, than all of its consecutive states. Let

$$u = f(x, y), \qquad \text{then} \qquad u' = f(x + h, y + k),$$

$$u' - u = h(p + p't) + \frac{h^2}{1.2}(q + 2q't + q''t^2) + \&c. \ldots\ldots (1),$$

after placing in the development of article (49),

$$k = ht, \qquad \frac{du}{dx} = p, \qquad \frac{du}{dy} = p',$$

$$\frac{d^2u}{dx^2} = q, \qquad \frac{d^2u}{dxdy} = q', \qquad \frac{d^2u}{dy^2} = q'' \;\&c.$$

The sign of this series, when h is infinitely small, will depend upon the sign of its first term. Now we shall obtain all of the consecutive states of u, by giving to h and k proper infinitely small values, both positive and negative; and therefore, when u is either a maximum or a minimum, the sign of $u' - u$ for all these values of h and k must be the same: But the first term of the series (1) evidently changes its sign when the sign of h changes; there can, then, be neither a maximum nor a minimum, unless

$$h(p + p't) = 0 \qquad \text{or} \qquad p + p't = 0,$$

and since this must be 0 for all values of $t = \frac{k}{h}$, we must have separately $\quad p = 0 \quad$ and $\quad p' = 0, \quad$ or

$$\frac{du}{dx} = 0 \ldots\ldots(2) \qquad \frac{du}{dy} = 0 \ldots\ldots(3).$$

The values of x and y, deduced from these equations and substituted in the second term of series (1), (h and k being infinitely small,) should make it negative for a maximum and positive for a minimum. This term may be put under the form

$$\frac{h^2q''}{1.2}\left(\frac{q}{q''} + \frac{2q'}{q''}t + t^2\right),$$

which, if there be a maximum or minimum, must not change its sign for any value of t; but this requires that the roots of the equation

$$t^2 + 2\frac{q'}{q''}t + \frac{q}{q''} = 0$$

be either imaginary or equal; that is, that q and q'' have the same sign, and $\quad q'^2 < qq'' \quad$ or $\quad q'^2 = qq''$.

The conditions then are

$$\left(\frac{d^2u}{dxdy}\right)^2 < \frac{d^2u}{dx^2} \times \frac{d^2u}{dy^2} \text{ or } = \frac{d^2u}{dx^2} \times \frac{d^2u}{dy^2};$$

and also that $\frac{d^2u}{dx^2}$ and $\frac{d^2u}{dy^2}$ have the same sign, after the values of x and y deduced from the equations $\frac{du}{dx} = 0$ and $\frac{du}{dy} = 0$ have been substituted: And since the sign of the second term will then depend upon q'', the sign of $\frac{d^2u}{dy^2}$ must be negative for a maximum, and positive for a minimum.

If the second term becomes 0, we must substitute the values of x and y in the third, which must also be 0, and the sign of the fourth negative for a maximum, and positive for a minimum; the discussion of the several conditions of which, although complicated, may be made in a manner similar to the above.

Examples.

1. Required to divide a number a into three parts, such that the cube of the first, into the square of the second, into the first power of the third, shall be a maximum.

Let $x =$ the first part, and $y =$ the second; then $a - x - y$ $=$ the third, and

$$u = x^3y^2(a - x - y),$$

$$\frac{du}{dx} = x^2y^2(3a - 3y - 4x), \qquad \frac{du}{dy} = x^3y(2a - 3y - 2x).$$

Placing these equal to 0, we have

$$3a - 3y - 4x = 0, \qquad 2a - 3y - 2x = 0;$$

whence

$$x = \frac{a}{2}, \qquad y = \frac{a}{3}.$$

We have also

$$q = \frac{d^2u}{dx^2} = 2xy^2(3a - 3y - 6x),$$

$$q' = \frac{d^2u}{dxdy} = x^2y(6a - 9y - 8x),$$

$$q'' = \frac{d^2u}{dy^2} = x^3(2a - 6y - 2x),$$

which for the particular values of x and y become

$$-\frac{a^4}{9}, \qquad -\frac{a^4}{12}, \qquad -\frac{a^4}{8}.$$

Hence

$$q'^2 = \frac{a^8}{144} < qq'' = \frac{a^8}{72}, \quad \text{and} \quad \frac{d^2u}{dy^2} = -\frac{a^4}{8};$$

u is therefore a maximum when its value is $\frac{a^6}{432}$.

2. Make the preceding proposition general, by putting for the cube, square, and first power, the mth, nth, and rth powers.

Then

$$u = x^m y^n (a - x - y)^r,$$

$$x = \frac{ma}{m + n + r}, \qquad y = \frac{na}{m + n + r}.$$

3. Required the shortest distance from a given point to a given plane.

Let the equation of the plane be placed under the form

$$z = \text{A}x + \text{B}y + \text{D},$$

and the co-ordinates of the given point be x', y', and z'; then

$$u = \sqrt{(x - x')^2 + (y - y')^2 + (z - z')^2},$$

or putting for z its value,

$$u = \sqrt{(x - x')^2 + (y - y')^2 + (\text{A}x + \text{B}y + \text{D} - z')^2}.$$

Calling the radical, R, we shall have

$$\frac{du}{dy} = \frac{y - y' + (\mathrm{A}x + \mathrm{B}y + \mathrm{D} - z')\mathrm{B}}{\mathrm{R}},$$

$$\frac{du}{dx} = \frac{x - x' + (\mathrm{A}x + \mathrm{B}y + \mathrm{D} - z')\mathrm{A}}{\mathrm{R}}.$$

Placing these equal to 0, and solving the resulting equations, we may obtain the values of x and y; and thence, of z. Or otherwise, putting for $\mathrm{A}x + \mathrm{B}y + \mathrm{D}$ its value z, we have

$$y - y' + \mathrm{B}(z - z') = 0, \quad \text{and} \quad x - x' + \mathrm{A}(z - z') = 0,$$

which are evidently the equations of a perpendicular to the plane, and if combined with the equation of the plane will give the values of x, y, and z.

75. In order that a function of three or more variables be a maximum or a minimum, we must have

$$\frac{du}{dx} = 0, \qquad \frac{du}{dy} = 0, \qquad \frac{du}{dz} = 0 \ldots\ldots\ldots\ldots \&c.,$$

and the relation between the partial differential coefficients of the second order must be such, that the second term, in the development of the difference $u' - u$ shall remain of the same sign, for all the consecutive values of the function.

APPLICATION OF THE DIFFERENTIAL CALCULUS TO CURVES.

76. To x in the expression

$$u = f(x),$$

assign a particular value, and deduce the corresponding value of u. These values, taken together, may be considered the co-ordinates of a point which may then be constructed. By assigning an infinite number of values to x, and deducing the corresponding values of u, an infinite number of points may be determined, which, being joined, will form a curve whose equation is $u = f(x)$. Hence, we conclude that *every function of a single variable may be regarded as the ordinate of a curve, of which the variable is the abscissa.*

77. Let BMM′ be a curve, the equation of which is $y = f(x)$; and M any point of this curve, the co-ordinates being x and y. Increase the abscissa AP or x, by the variable increment PP′ $= h$; denote the corresponding ordinate P′M′ by y'; and draw the secant M′MT′. Then

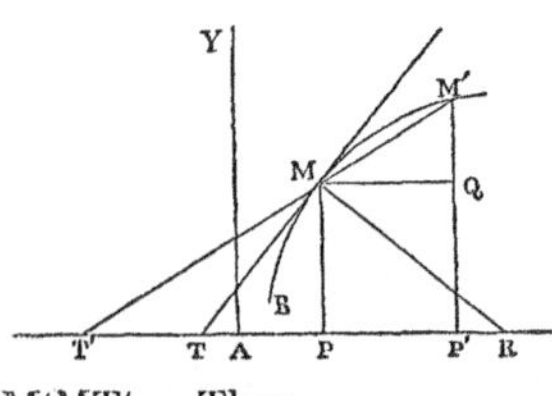

$$M'Q = P'M' - PM = y' - y = Ph + P'h^2 \text{......Art. (10).}$$

From the triangle M′MQ, we have

$$\text{tang } M'MQ = \frac{M'Q}{MQ} = \text{tang } MT'X,$$

and placing for M′Q and MQ $=$ PP′, their values, this becomes

$$\text{tang } MT'X = \frac{Ph + P'h^2}{h} = P + P'h \text{......(1).}$$

Now if h be diminished, the point M′ approaches M, and the secant M′T′ approaches the tangent MT, and finally when $h = 0$, the point M′ coincides with M, and the secant with the tangent. If then in (1) we make $h = 0$, we have

$$\text{tang MTX} = P = \frac{dy}{dx};$$

that is, *the tangent of the angle which a tangent line at any point of a curve makes with the axis of X, is equal to the first differential coefficient of the ordinate of the curve.* To show the application of this principle, let us take the equation of a circle

$$x^2 + y^2 = R^2;$$

whence

$$\frac{dy}{dx} = -\frac{x}{y} \ldots\ldots\ldots\ldots(2);$$

for the general value of the tangent of the angle made by a tangent at any point of the circumference, with the axis of X.

If the particular value at a point whose co-ordinates are x'' and y'' be required; for x and y, let x'' and y'' be substituted, then

$$\frac{dy''}{dx''} = -\frac{x''}{y''}.*$$

Take also the equation

$$y^2 = mx + nx^2;$$

whence

$$\frac{dy}{dx} = \frac{m + 2nx}{2y} = \frac{m + 2nx}{2\sqrt{mx + nx^2}}.$$

For the particular point y'' and x'', this expression becomes

---

* NOTE.—The notation $\frac{dy''}{dx''}$, $\frac{d^2y''}{dx''^2}$, &c., is used to indicate what the first, second, &c. differential coefficients become, when for the general variables x and y the particular values x'' and y'' are substituted.

$$\frac{dy''}{dx''} = \frac{m + 2nx''}{2\sqrt{mx'' + nx''^2}}.$$

78. If it be required to find the point of a given curve, at which the tangent line makes a given angle with the axis of X, we know that at this point the first differential coefficient must be equal to the tangent of the given angle. Calling this tangent a, we must then have

$$\frac{dy}{dx} = a,$$

and this combined with the equation of the curve will give the particular values of x and y, for the required point.

If the tangent line is to be parallel to the axis of X, then for the point of tangency, $\frac{dy}{dx} = 0$; and if perpendicular, $\frac{dy}{dx} = \infty$.

We will illustrate each of these cases by an example.

1. Let it be required to find the point on a given parabola, at which the tangent line makes an angle of 45° with the axis. The equation of the parabola is $y^2 = 2px$, by the differentiation of which, &c. we have

$$\frac{dy}{dx} = \frac{p}{y}.$$

But as tang 45° $= 1$, we have, for the required point,

$$\frac{dy}{dx} = \frac{p}{y} = 1,$$

and, combining this with the equation $y^2 = 2px$, we find

$$x = \frac{p}{2} \qquad y = p.$$

The tangent at the extremity of the ordinate passing through the focus, will then fulfil the required condition.

2. Let

$$y = a + (c - x)^2 \dots\dots\dots\dots(1)$$

represent a curve; then

$$\frac{dy}{dx} = - 2(c - x),$$

which is equal to 0, when $x = c$; and this value of x in (1) gives $y = a$. These are then the co-ordinates of the point at which the tangent is parallel to the axis of X.

3. Let

$$y = a + (c - x)^{\frac{1}{2}}$$

represent a curve; then

$$\frac{dy}{dx} = - \frac{1}{2(c - x)^{\frac{1}{2}}}$$

which is equal to infinity, when $x = c$. $\therefore x = c$ and $y = a$ are then the co-ordinates of the point at which the tangent is perpendicular to the axis of X.

79. If x'' and y'' represent the co-ordinates of a given point on a given curve, whose equation is $y = f(x)$; the equation of a straight line passing through this point will be

$$y - y'' = a(x - x''),$$

a being indeterminate. This will become *the equation of a tan-*

gent line at the given point, if for a we put $\frac{dy''}{dx''}$. We thus obtain

$$y - y'' = \frac{dy''}{dx''}(x - x'')\text{............}(1).$$

By differentiating the equation of an ellipse

$$a^2y^2 + b^2x^2 = a^2b^2,$$

we deduce

$$\frac{dy}{dx} = -\frac{b^2x}{a^2y}; \qquad \text{whence} \qquad \frac{dy''}{dx''} = -\frac{b^2x''}{a^2y''},$$

and this value in (1) gives, for the equation of a tangent, to an ellipse at the point y'', x'',

$$y - y'' = -\frac{b^2x''}{a^2y''}(x - x'')$$

which, by reduction, becomes $a^2yy'' + b^2xx'' = a^2b^2$.

80. If the equation of a tangent be required, which shall be parallel to a given line, or make a given angle with the axis of X; we may determine the co-ordinates of the point of contact as in article (78); and knowing these, the equation may be deduced as above.

Thus, if a tangent to a circle be required to make with the axis of X an angle whose tangent is 2, we must have for the required point, equation (2), Art. (77),

$$\frac{dy}{dx} = -\frac{x}{y} = 2.$$

From this, we find $y = -\frac{x}{2}$, which, combined with the equation of the circle, gives

$$x = \pm \frac{2R}{\sqrt{5}} = x'' \qquad y = \mp \frac{R}{\sqrt{5}} = y''.$$

and equation (1), Art. (79), becomes, when we use the upper signs,

$$y + \frac{R}{\sqrt{5}} = 2\left(x - \frac{2R}{\sqrt{5}}\right) \quad \text{or} \quad y = 2x - R\sqrt{5}.$$

81. The general equation of a normal, deduced from equation (1), article (79), is evidently

$$y - y'' = -\frac{1}{\frac{dy''}{dx''}}(x - x''),$$

or

$$y - y'' = -\frac{dx''}{dy''}(x - x'').$$

82. The right angled triangle MTP (Figure of Art. 77) gives
PM = PT tang MTP; hence $\text{PT} = \frac{\text{PM}}{\text{tang MTP}}$;

or

$$\textit{Subtangent} = \frac{y}{\frac{dy}{dx}} = y\frac{dx}{dy}.$$

Also

$$\text{MT} = \sqrt{\overline{\text{MP}}^2 + \overline{\text{PT}}^2},$$

or

$$Tangent = \sqrt{y^2 + y^2\frac{dx^2}{dy^2}} = y\sqrt{1 + \frac{dx^2}{dy^2}}.$$

The right-angled triangle PMR gives

$$\text{PR} = \text{MP tang PMR}; \qquad \text{but} \qquad \text{PMR} = \text{MTP};$$

hence

$$\text{PR} = \text{MP tang MTP}, \qquad \text{or} \qquad Subnormal = y\frac{dy}{dx}.$$

Also,

$$\text{MR} = \sqrt{\overline{\text{MP}}^2 + \overline{\text{PR}}^2};$$

hence

$$Normal = \sqrt{y^2 + y^2\frac{dy^2}{dx^2}} = y\sqrt{1 + \frac{dy^2}{dx^2}}.$$

To apply these formulas to a particular curve, it is only necessary to substitute in each the value of $\frac{dx}{dy}$, or $\frac{dy}{dx}$, deduced from the differential equation of the curve. The results will be general for all points of the curve. If the values for a given point be required, in these results let the co-ordinates of the point be substituted for x and y.

For example, take the general equation of Conic Sections,

$$y^2 = mx + nx^2;$$

whence

$$\frac{dy}{dx} = \frac{m + 2nx}{2\sqrt{mx + nx^2}} \qquad \frac{dx}{dy} = \frac{2\sqrt{mx + nx^2}}{m + 2nx}.$$

These values substituted in the formulas, give

$$PT = \frac{2(mx + nx^2)}{m + 2nx}. \qquad PR = \frac{m + 2nx}{2}.$$

$$MT = \sqrt{mx + nx^2 + 4\left(\frac{mx + nx^2}{m + 2nx}\right)^2}.$$

$$MR = \sqrt{mx + nx^2 + \frac{1}{4}(m + 2nx)^2}.$$

For the parabola $n = 0$, and these expressions become

$$PT = 2x. \qquad PR = \frac{m}{2}.$$

$$MT = \sqrt{mx + 4x^2}. \qquad MR = \sqrt{mx + \frac{m^2}{4}}.$$

83. If a curve be convex towards the axis of X, and the ordinate *positive*, as in the annexed figure, it is plain, that as the abscissas AP, AP′, &c. increase, the tangents of the angles MTX, M′T′X, &c., will also increase, and the reverse. Since these tangents are represented by the corresponding values of the first differential co-efficient of the ordinate $\left(\frac{dy}{dx}\right)$, *it must be an increasing function of* x, and its differential coefficient, i. e., $\frac{d^2y}{dx^2}$, must be *positive*, Art. (12).

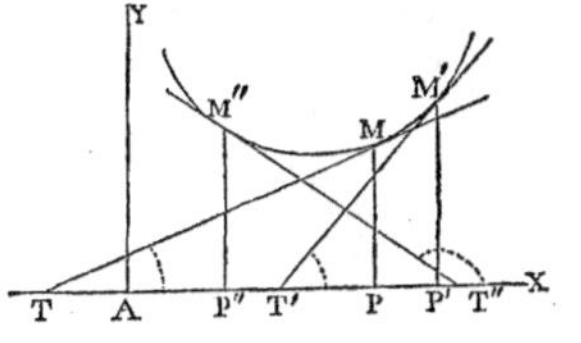

If the curve be still convex, and the ordinate *negative*; the tangents of the angles STX, S′T′X, &c. plainly decrease as x increases; $\frac{dy}{dx}$

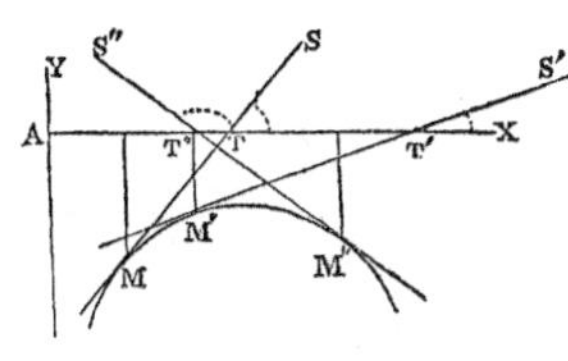

is a decreasing function of x; and $\frac{d^2y}{dx^2}$ must be *negative.*

If then a curve be convex towards the axis of abscissas, *the ordinate and its second differential coefficient, taken at the different points, will have the same sign.*

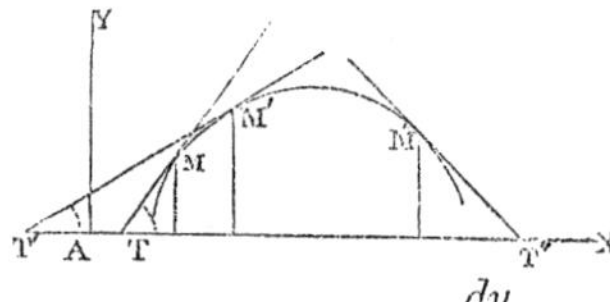

If the curve be concave, and the ordinate *positive*, as in the figure; the tangents of the angles MTX, M′T′X, &c. will decrease as x increases; $\frac{dy}{dx}$ will be a decreasing function, and $\frac{d^2y}{dx^2}$ *negative*

If the curve be concave and the ordinate *negative;* the reverse will evidently be the case, and $\frac{d^2y}{dx^2}$ will be *positive.*

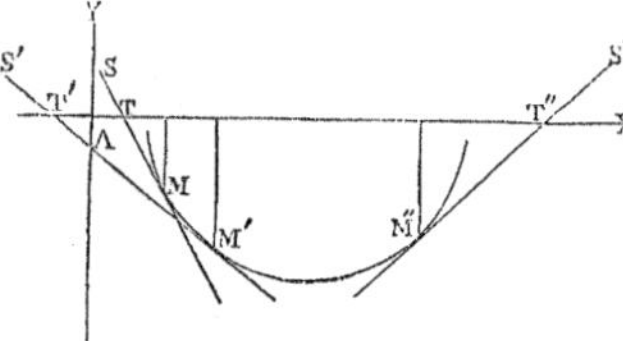

Hence if a curve be concave towards the axis of abscissas, *the ordinate and its second differential coefficient will have contrary signs.*

ASYMPTOTES.

84. An asymptote is a line which, continually approaching a curve, becomes tangent to it at an infinite distance. Asymptotes may be *curvilinear* or *rectilinear.* The latter only will be considered here.

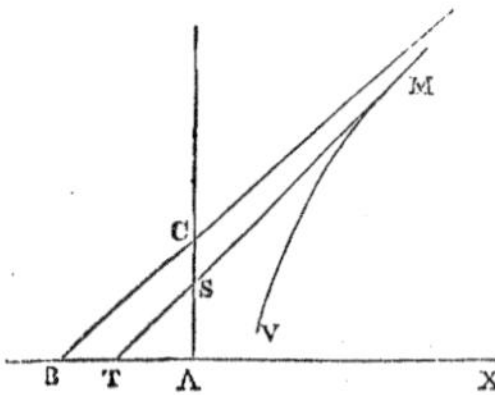

Let MV be any curve, and BC a rectilinear asymptote. Also let MT be any tangent line, the equation of which, article (79), is

$$y - y'' = \frac{dy''}{dx''}(x - x'').$$

If we make $y = 0$ in this equation, we obtain

$$x = x'' - y''\frac{dx''}{dy''} = \text{AT}\ldots\ldots\ldots\ldots(1).$$

If we make $x = 0$, we obtain

$$y = y'' - x''\frac{dy''}{dx''} = \text{AS}\ldots\ldots\ldots\ldots(2).$$

Now as the point of contact M, the co-ordinates of which are x'' and y'', is removed farther from the origin, the tangent MT will approach nearer to the asymptote BC; and finally, when M is at an infinite distance, the two will coincide, and the distances AT and AS become respectively equal to the distances AB and AC.

If then the expressions for AT and AS, *when such values are substituted for x'' and y'' as to remove the point M to an infinite distance*, are both finite, there will be an asymptote, which may be drawn through the points B and C.

If one of these expressions becomes infinite, and the other finite, there will be an asymptote parallel to the axis on which the distance is infinite.

If both expressions become infinite, there will be no asymptote. If both become 0, the asymptote will pass through the origin of co-ordinates, and the tangent of the angle which it makes with the axis of X may be obtained from the value of $\frac{dy''}{dx''}$, when for x'' and y'' the proper values are substituted.

Hence to construct the asymptote of a given curve; find, by differentiating the equation of the curve, the values of $\frac{dy''}{dx''}$ and $\frac{dx''}{dy''}$, which substitute in formulas (1) and (2); the results thence obtained by substituting for x'' and y'' their values for that point of the curve which is at an infinite distance, will be the distances cut off from the co-ordinate axes by the asymptote, if there is one.

Examples.

1. Take the equation of lines of the second order,

$$y^2 = mx + nx^2.$$

By differentiation, &c., we obtain

$$\frac{dy}{dx} = \frac{m + 2nx}{2y} = \frac{m + 2nx}{\pm 2\sqrt{mx + nx^2}};$$

whence

$$\frac{dy''}{dx''} = \frac{m + 2nx''}{\pm 2\sqrt{mx'' + nx''^2}}.$$

Substituting this and the value of $y'' = \pm \sqrt{mx'' + nx''^2}$ in (1) and (2), we have

$$\text{AT} = x'' - \frac{2(mx'' + nx''^2)}{m + 2nx''} = \frac{-mx''}{m + 2nx''} = \frac{-m}{\frac{m}{x''} + 2n} \ldots\ldots(3).$$

$$\text{AS} = \pm\sqrt{mx'' + nx''^2} - \frac{mx'' + 2nx''^2}{\pm 2\sqrt{mx'' + nx''^2}} = \frac{m}{\pm 2\sqrt{\frac{m}{x''} + n}} \ldots\ldots(4).$$

In this case, the co-ordinates of that point of the curve which is at an infinite distance are $x'' = \infty$ and $y'' = \infty$. Making $x'' = \infty$ in (3) and (4), we have

$$AB = -\frac{m}{\frac{m}{\infty} + 2n} = -\frac{m}{2n}.$$

$$AC = \frac{m}{\pm 2\sqrt{\frac{m}{\infty} + n}} = \frac{m}{\pm 2\sqrt{n}}.$$

For the hyperbola n is positive, these expressions are both finite, and, as there are two different values of AC, there are two asymptotes, and since $m = \frac{2B^2}{A}$ and $n = \frac{B^2}{A^2}$, we have

$$AB = -A, \qquad AC = \pm B.$$

For the parabola $n = 0$, the expressions are both infinite, and there is no asymptote.

For the ellipse n is negative and AC imaginary, as it should be, since there is no point of the curve at an infinite distance, and of course no asymptote.

2. Take the equation

$$x^3 - 3axy + y^3 = 0.$$

From formulas (1) and (2), we obtain in this case

$$AT = \frac{ax''y''}{x''^2 - ay''} \ldots\ldots\ldots\ldots(5), \qquad AS = \frac{ax''y''}{y''^2 - ax''} \ldots\ldots\ldots\ldots(6).$$

As it is difficult to obtain the value of y in terms of x from the given equation, we can not at once eliminate y'' from (5) and (6); but if we make $x = ty$ and substitute in the given equation, it will be divisible by y^2, and we then find

$$y = \frac{3at}{1 + t^3}.$$

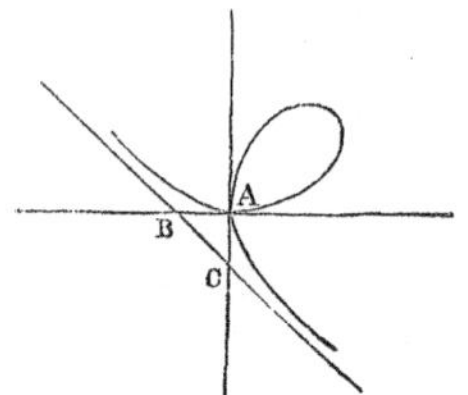

If in this we make $t = -1$, we have $y = \infty$; hence when y is infinite it is equal to $-x$, and for that point of the curve which is at an infinite distance we have $y'' = -x'' = \infty$.

Changing y'' into $-x''$ in (5) and (6), they become

$$AT = \frac{-a}{1 + \frac{a}{x''}}, \qquad AS = \frac{-a}{1 - \frac{a}{x''}};$$

and making $x'' = \infty$, we find

$$AB = -a \qquad \text{and} \qquad AC = -a;$$

hence BC is the asymptote.

3. Take the equation $\qquad x^2y^3 = p,$
in the curve represented by which, the points at an infinite distance have for their co-ordinates $x'' = 0$, $y'' = \infty$, and $y'' = 0$, $x'' = \infty$.

DIFFERENTIALS OF AN ARC, AREA, &C.

85. Let u represent any function of x and

$$Q \qquad \text{and} \qquad Q'$$

two functions of x and h which have the same limit, denoted by m; and suppose that for all small values of h

$$Q < \frac{u' - u}{h} < Q'.$$

Since Q and Q′ have the same limit m, they must approach an equality as h diminishes, and each reduce to m when h becomes *zero*; and since $\frac{u'-u}{h}$ cannot for any small value of h be less than Q nor greater than Q′, it follows, that being thus comprehended between Q and Q′ it must with them, as h diminishes, approach nearer and nearer to m, and when h becomes 0, it must also reduce to m, or *m must be its limit*, that is, Art. (7),

$$\frac{du}{dx} = m.$$

Hence; *if the ratio of the increment* (h) *of the variable to that of the function, for all small values of h, be greater than a certain quantity, the limit of which is m, and at the same time less than a certain other quantity the limit of which is m, then will the differential coefficient of the function be equal to m.*

We must therefore have

$$u' - u = mh + P'h^2 \ldots\ldots\ldots\ldots \text{Art. (10).}$$

By dividing this, first by Qh and then by $Q'h$, we have

$$\frac{m + P'h}{Q} \qquad\qquad \frac{m + P'h}{Q'}$$

the limit of either of which is $\frac{m}{m} = 1$: *Hence the limit of the ratio obtained by dividing the increment of the function by either of the quantities into h, is unity.*

86. Let BM $= z$ be any arc of a curve, the equation of which is $y = f(x)$. Although z changes whenever x or y is changed, yet the equation $y = f(x)$ establishes such a relation between x and y, that one is necessarily a function of the other. z may

therefore be regarded as a function of either. Let us regard it as a function of x, and let $AP = x$, $PM = y$; and increase x by

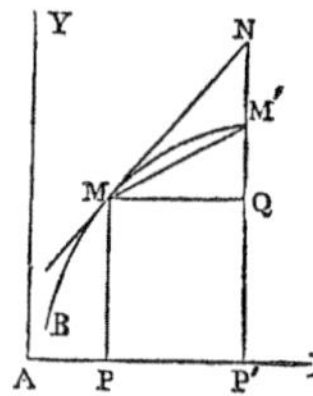

$PP' = h$; then

$$BM' = z', \qquad \text{and} \qquad MM' = z' - z$$

will be the increment of the arc z, and

$$M'Q = y' - y = Ph + P'h^2,$$

the increment of the ordinate y.

Draw the tangent MN at the point M. We then have

$$\text{tang NMQ} = \frac{dy}{dx} = P,$$

$$NQ = Ph, \qquad NQ - M'Q = NM' = -P'h^2,$$

$$MM' = \sqrt{\overline{MQ}^2 + \overline{M'Q}^2} = h\sqrt{1 + (P + P'h)^2},$$

$$MN = \sqrt{\overline{MQ}^2 + \overline{NQ}^2} = h\sqrt{1 + P^2},$$

$$MN + NM' = h\sqrt{1 + P^2} - P'h^2.$$

But the arc MM' is greater than the chord MM', and less than the broken line $MN + NM'$ for all values of h; therefore

$$z' - z > h\sqrt{1 + (P + P'h)^2}, \quad \text{and} \quad z' - z < h\sqrt{1 + P^2} - P'h^2$$

or

$$\frac{z' - z}{h} > \sqrt{1 + (P + P'h)^2}, \qquad \frac{z' - z}{h} < \sqrt{1 + P^2} - P'h;$$

the second members of each of which expressions become

$$\sqrt{1 + P^2}, \qquad \text{when} \qquad h = 0.$$

Therefore, in accordance with the principle of the preceding article, we must have

$$\frac{dz}{dx} = \sqrt{1 + P^2} = \sqrt{1 + \frac{dy^2}{dx^2}}$$

$$dz = \sqrt{dx^2 + dy^2};$$

that is, *the differential of an arc is equal to the square root of the sum of the squares of the differentials of the co-ordinates of its points.*

To illustrate, take the equation of a circle

$$x^2 + y^2 = R^2;$$

whence

$$dy = -\frac{xdx}{y} = -\frac{xdx}{\sqrt{R^2 - x^2}},$$

and

$$dz = \sqrt{dx^2 + \frac{x^2dx^2}{R^2 - x^2}} = \frac{Rdx}{\sqrt{R^2 - x^2}}.$$

87. Since, also, by article (85), the limit of the ratio

$$\frac{z' - z}{h\sqrt{1 + (P + P'h)^2}} = \frac{z' - z}{MM'}$$

is unity, we prove that *the limit of the ratio of a chord to its corresponding arc is unity.*

88. Let BMP $= s$, be any area limited by a curve and the axis of X; it will evidently be a function of x.

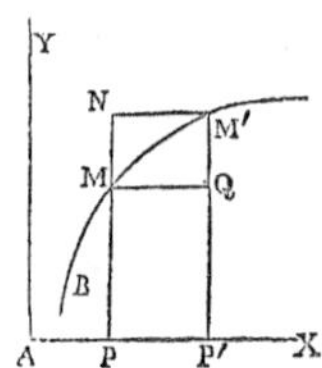

Let $AP = x, \quad PM = y, \quad PP' = h;$

then $P'M' = y + Ph + P'h^2,$

and $PMM'P' = s' - s$

the increment of the area s.

The rectangles PM′ and PQ being constructed, we have

$$PQ = yh, \qquad PM' = P'M'h = (y + Ph + P'h^2)h.$$

But the area PMM′P′ is always greater than the rectangle PQ, and less than PM′; whence

$$s' - s > yh, \qquad \text{and} \qquad s' - s < (y + Ph + P'h^2)h,$$

or

$$\frac{s' - s}{h} > y, \qquad \frac{s' - s}{h} < y + Ph + P'h^2,$$

both of which quantities become y, when $h = 0$; hence, Art. (85),

$$\frac{ds}{dx} = y \qquad \text{and} \qquad ds = ydx;$$

that is, *the differential of the area is equal to the ordinate of the bounding curve into the differential of the abscissa.*

The differential of the area included between the curve and axis of Y, may be found in the same way to be

$$ds = xdy.$$

If the axes of co-ordinates are oblique to each other, the rectangles PQ and PM′ become parallelograms; the area of the first is

$$yh\sin\omega,$$

and of the second

$$(y + Ph + P'h^2)h\sin\omega,$$

ω representing the angle made by the axes of co-ordinates; whence

$$ds = y \sin \omega \, dx.$$

For an example, take the equation

$$y^2 = R^2 - x^2;$$

whence

$$ds = ydx = \sqrt{R^2 - x^2}dx.$$

Take also the equation

$$y^2 = 2p'x,$$

the axes of co-ordinates being oblique; then

$$ds = \sqrt{2p'x} \sin \omega \, dx.$$

89. Let the curve BM revolve about the axis of X; it will generate a surface of revolution which will be a function of x, and which we denote by u.

The notation being as in the preceding articles, the increment $(u' - u)$ of the surface, when x is increased by h, will be generated by the arc MM'. The line MN, tangent at M, generates the surface of a frustrum of a cone which has for its measure, (since $P'N = P'Q + NQ = y + Ph$),

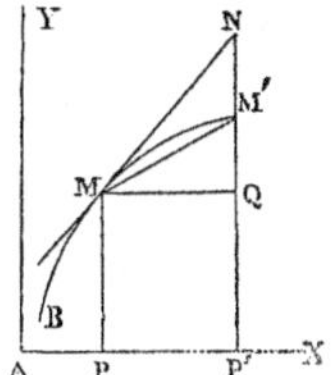

$$2\pi(PM + P'N)\frac{MN}{2} = \pi(2y + Ph)\sqrt{h^2 + P^2h^2}.$$

The line M'N generates a plane surface which is equal to the difference between the two circles whose radii are P'N and P'M', that is,

$$\pi(y + Ph)^2 - \pi(y + Ph + P'h^2)^2 = Rh^2,$$

expressing by R the polynomial coefficient of h^2, after reduction. The sum of these two surfaces always exceeds the surface $u' - u$, therefore

$$\frac{u' - u}{h} < \pi(2y + Ph)\sqrt{1 + P^2} + Rh.$$

The chord MM′ generates the frustrum of a cone which is less than $u' - u$ and measured by

$$2\pi(PM + P'M')\frac{MM'}{2} = \pi(2y + Ph + P'h^2)\sqrt{h^2 + (Ph + P'h^2)^2};$$

hence

$$\frac{u' - u}{h} > \pi(2y + Ph + P'h^2)\sqrt{1 + (P + P'h)^2}.$$

Since the second members of both the above inequalities reduce to $2\pi y\sqrt{1 + P^2}$ when $h = 0$, we must have

$$\frac{du}{dx} = 2\pi y\sqrt{1 + P^2} = 2\pi y\sqrt{1 + \frac{dy^2}{dx^2}},$$

$$du = 2\pi y\sqrt{dx^2 + dy^2};$$

that is, the differential of a surface of revolution *is equal to the circumference of a circle perpendicular to the axis, multiplied by the differential of the arc of the generating curve.* If the curve revolve about the axis of Y, we may determine in the same way

$$du = 2\pi x\sqrt{dx^2 + dy^2}.$$

If we suppose a parabola, whose equation is $y^2 = 2px$, to revolve about its axis, we shall have

$$dx = \frac{ydy}{p},$$

$$du = 2\pi y\sqrt{\frac{y^2dy^2}{p^2} + dy^2} = \frac{2\pi ydy}{p}\sqrt{y^2 + p^2}.$$

90. Let the area BMP revolve about the axis of X; it will generate a solid of revolution, which is a function of x, and which we denote by v. If x be increased by $PP' = h$, then the area PMM′P′ will generate the increment $(v' - v)$ of the solid. The rectangle PQ will generate a cylinder, which is always less than $v' - v$, and which is measured by $\pi y^2 h$. The rectangle PM′ will generate another cylinder, which is always greater than $v' - v$, and is measured by $\pi(y + Ph + P'h^2)^2 h$; hence we have

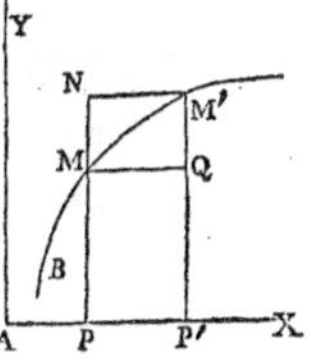

$$\frac{v' - v}{h} > \pi y^2 \qquad \text{and} \qquad \frac{v' - v}{h} < \pi(y + Ph + P'h^2)^2,$$

therefore

$$\frac{dv}{dx} = \pi y^2 \qquad\qquad dv = \pi y^2 dx;$$

that is, the differential of a solid of revolution *is equal to the area of a circle perpendicular to the axis, multiplied by the differential of the abscissa of the curve which generates the bounding surface.*

For the solid generated by the area included between the curve and axis of Y, we may find in the same way,

$$dv = \pi x^2 dy.$$

If we take the particular case of the ellipsoid, the equation of the generating curve being

$$y^2 = \frac{b^2}{a^2}(a^2 - x^2),$$

we have

$$dv = \pi y^2 dx = \pi \frac{b^2}{a^2}(a^2 - x^2)dx.$$

GENERAL REMARKS.

91. Heretofore, in our treatment of the subject, we have regarded the differential of the independent variable merely as an arbitrary constant, Art. (7), without having fixed upon any particular value for it. All the demonstrations are then as true for one value, as for another.

It is however of the greatest convenience, in the application of the Calculus to the higher branches of Mathematics and Physical Science, to regard this differential as *infinitely small*; that is, *so small as to be contained in unity an infinite number of times*; and hereafter it will be so regarded.

The advantages of so regarding it will appear evident after a few illustrations. Let us take first the simple function discussed in article (7),

$$u = ax^2.$$

After x has been increased by dx, we have there found

$$u' - u = 2axdx + adx^2.$$

Now, if the increment (dx) of the variable be infinitely small, the two states u and u' will plainly be consecutive, the expression for their difference being

$$2axdx + adx^2 \ldots\ldots\ldots\ldots(1).$$

But since dx is infinitely small, its square will be infinitely small when compared with it: As may be shown by taking the identical equation

$$\frac{1}{dx} = \frac{dx}{dx^2} = \frac{dx^2}{dx^3} = \&c.,$$

from which, since dx is contained an infinite number of times in unity, it appears that dx^2 will be contained an infinite number of times in dx; dx^3 in dx^2, &c.: adx^2 will then be infinitely small with reference to $2axdx$, and may be omitted from expression (1) without materially affecting its value; hence in this case $2axdx$ *may be taken for, or is the measure of, the difference* $u' - u$.

This is true also in the general case, for all the terms of the difference, except the one which we have taken for the differential, will contain dx to a higher power than the first [see equation (3), Art. (7)]; they may then all be rejected, and the differential of the function taken, as *the measure of the difference between two consecutive states of the function.*

It is plain, also, since

$$du = pdx, \qquad d^2u = qdx^2, \qquad d^3u = rdx^3, \text{ \&c.......Art. (26)},$$

that the second differential of a function is infinitely small when compared with the first, and the third when compared with the second, and so on. It is usual to call these, infinitely small quantities of the first, second, and third orders; and we see, from what precedes, that *every infinitely small quantity may be omitted without error, when connected by the sign* $\pm$ *with one of a lower order.*

In the application of the Calculus to curves these principles are of great use. Let BMM′ be a curve; MP, M′P′, any two consecutive ordinates; PP′ = P′P″ = P″P‴, &c., being each equal to dx; then the difference between y and y', or $y' - y$ = M′Q, is equal to dy; and $z' - z$ = MM′ = dz:

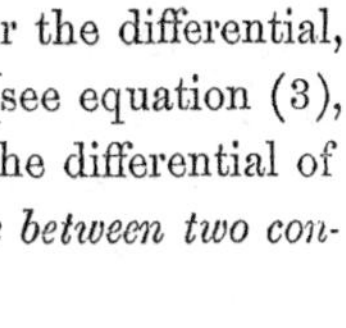

Or since $z' - z$ may represent the difference MM′, M′M″, M″M‴, between any two consecutive states of the arc, the different values of dz may in succession represent the infinitely small arcs MM′, M′M″, &c. the sum of all of which will be equal to the entire arc z.

So the difference between the two areas BM′P′ and BMP is equal to PMM′P′ $= ds$; and the different values of ds may in succession represent the infinitely small areas PMM′P′, P′M′M″P″, &c., the sum of all of which will equal the entire area s: And in general, if the variable be increased by its differential, *the corresponding increment of the function may be represented by the differential of the function, and the sum of all the different values of this differential will equal the function itself.*

In accordance with the above principles, the expressions in articles (86), (88), (89) and (90) are at once deduced.

1. The arc MM′ is equal to $z' - z = dz$; and since the limit of the ratio of the arc and chord is unity, they continually approach an equality as they decrease; and when both are infinitely small, the one may be taken for the other. But

$$\text{the chord MM}' = \sqrt{\overline{\text{MQ}}^2 + \overline{\text{M}'\text{Q}}^2} = \sqrt{dx^2 + dy^2};$$

hence

$$dz = \sqrt{dx^2 + dy^2}.$$

And if x, y and z denote the co-ordinates of the points of a curve w in space, we may find in a similar way

$$dw = \sqrt{dx^2 + dy^2 + dz^2}.$$

2. The area PMM′P′ $= s' - s = ds$; and since the limit of the ratio $\dfrac{\text{PMM}'\text{P}'}{\text{PQ}}$ is unity, the area PMM′P′, when infinitely small, may be taken for the rectangle PQ. But

$$PQ = PM \times PP' = ydx;$$

hence

$$ds = ydx.$$

3. The surface generated by the arc MM′ is equal to $u' - u = du$, and this will not differ from the surface of the frustrum generated by the infinitely small chord MM′, which is equal to

$$[2\pi y + 2\pi(y + dy)] \frac{MM'}{2} = \pi(2y + dy)dz = 2\pi ydz,$$

since dy may be rejected; hence

$$du = 2\pi ydz = 2\pi y \sqrt{dx^2 + dy^2}.$$

4. The solid generated by the area PMM′P′ $= v' - v = dv$ will not differ from the solid generated by the rectangle PQ which is equal to

$$\pi\overline{MP}^2 \times PP' = \pi y^2 dx;$$

hence

$$dv = \pi y^2 dx.$$

SINGULAR POINTS.

92. *A singular point of a curve* is one at which there exists some remarkable property, not enjoyed by the other points.

By a general discussion of the equation

$$y = b + c(x - a)^m \text{......}(1),$$

we shall meet with some particular curves, on which some of these points will be found.

1st. *Let m be an entire and even number.*

By the differentiation, &c. of (1), we have

$$\frac{dy}{dx} = mc(x-a)^{m-1} \ldots\ldots(2), \quad \frac{d^2y}{dx^2} = m(m-1)c(x-a)^{m-2} \ldots\ldots(3),$$

. . . .

$$\frac{d^my}{dx^m} = m(m-1) \ldots\ldots 2.1.c.$$

Placing $\frac{dy}{dx} = 0$, we obtain $x = a$.

This value of x, when substituted in (1), (2), (3), &c., gives $y = b$, and reduces the successive differential coefficients to 0, as far as the mth, which, *if c be positive*, becomes a positive constant, and is of an even order; hence $y = b$ is a minimum ordinate, Art. (67).

Since for $x = a$, we have $\frac{dy}{dx} = 0$; the tangent line at the extremity of this minimum ordinate is parallel to the axis of X; and since (m and $m - 2$ being even) for all values of x except $x = a$, y and $\frac{d^2y}{dx^2}$ are positive, the curve at all of its points is convex towards the axis of X, Art. (83).

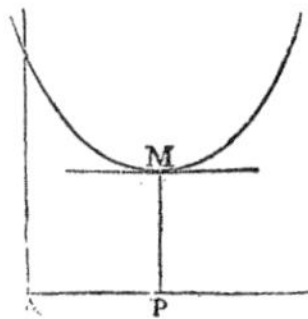

If c be negative; the mth differential coefficient will be negative; and $x = a$ and $y = b$ will be the co-ordinates of a point at which the ordinate is a maximum. In this case, the second differential coefficient for all values of x, except $x = a$, is negative, and the curve, for all positive values of y,

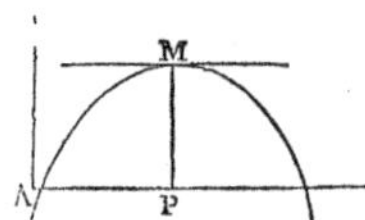

concave, and for all negative values of y, convex, towards the axis of X.

2d. *Let m be an entire and odd number.*

When $x = a$, the first differential coefficient as before, is equal to 0, as also the second, third, &c. The mth, however, *if c be positive*, is a positive constant, and of an odd order; there is then, in this case, neither a maximum nor a minimum, Art. (67).

By examining the second differential coefficient, we see (since $m - 2$ is odd), that for every value of $x < a$, it is negative; that for $x = a$, it is 0; and when $x > a$, it is positive; hence for all values of $x < a$, which give y positive, the curve is concave towards the axis of X; and for all values of $x > a$ it is convex, as in the figure.

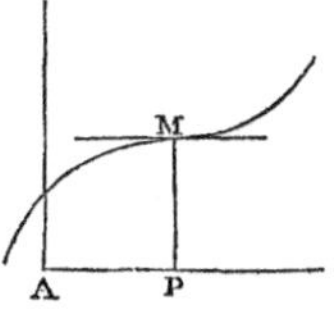

Therefore at the point whose co-ordinates are $x = a$ and $y = b$, as x increases, the curve changes from being concave, and becomes convex, towards the axis of X.

If c be negative; the reverse will be the case, and as in the second figure, at the point M, whose co-ordinates are $x = a$ and $y = b$, there is a change from convexity to concavity towards the axis of X. Such points are singular, and are called points of inflexion. In both cases the tangent line at the point of inflexion is parallel to the axis of X, and also cuts the curve.

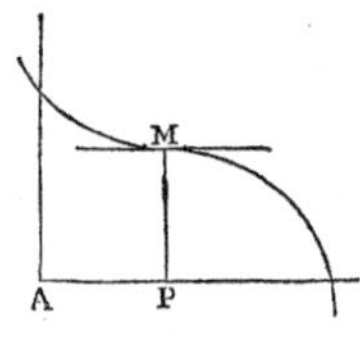

3d. *Let m be a fraction, the numerator and denominator of which are odd, as $\frac{3}{5}$.* Then

$$y = b + c(x - a)^{\frac{3}{5}},$$

$$\frac{dy}{dx} = \frac{3c}{5(x - a)^{\frac{2}{5}}}, \qquad \frac{d^2y}{dx^2} = -\frac{2}{5}\,\frac{3c}{5(x - a)^{\frac{7}{5}}} \ldots\ldots \&c.;$$

$x = a$ gives

$$y = b \qquad \frac{dy}{dx} = \infty \qquad \frac{d^2y}{dx^2} = \infty, \text{ \&c......}$$

If c be positive; $\frac{d^2y}{dx^2}$ for all values of $x < a$ will be positive, and

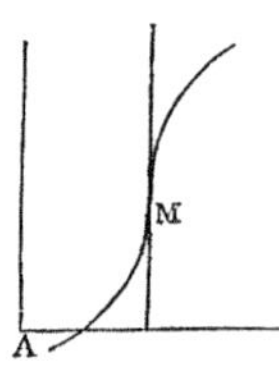

for all values of $x > a$, negative; hence for all values of x less than a which give y positive, the curve will be convex, and for all values of x greater than a it will be concave towards the axis of X, as in the figure.

If c be negative; the reverse is the case, as in the second figure.

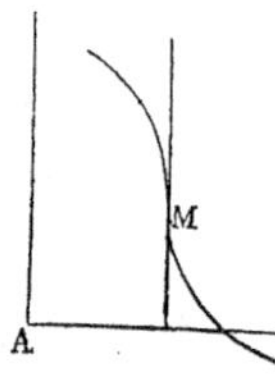

The point M, whose co-ordinates are $x = a$ and $y = b$, is in both cases a point of inflexion at which the tangent line is perpendicular to the axis of X. Whence we may say: *a point of inflexion is one at which, as the abscissa increases, a curve changes from being concave towards any right line, not passing through the point, and becomes convex, or the reverse.*

If the right line be taken as the axis of abscissas, this point will always be characterized by *a change of sign in the second differential coefficient of the ordinate.* For, since the curve on one side of the point is concave, and on the other convex, the second differential coefficient in one case has a different sign from that of the ordinate, and in the other the same; hence at the point the sign must have changed. In order that this may be the case, the second differential coefficient must be equal to zero, or infinity.

The roots of the two equations

$$\frac{d^2y}{dx^2} = 0 \qquad \text{and} \qquad \frac{d^2y}{dx^2} = \infty,$$

will then give all the values of the variable which can belong to points of inflexion.

It sometimes happens that a point of inflexion lies on the axis of X, as in the second case above discussed when $b = 0$. In this case $x = a$ gives

$$y = 0 \qquad \text{and} \qquad \frac{dy}{dx} = 0,$$

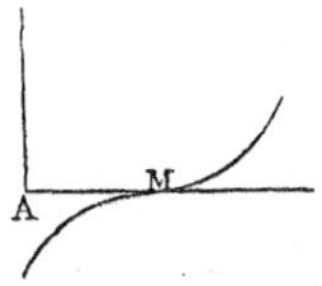

and the corresponding point M is a point of inflexion, at which both the second differential coefficient and ordinate change their signs.

It is evident from the preceding discussion, that if any right line be drawn through a point of inflexion, the curve on both sides of the point will either be convex towards the line, or concave.

4th. *Let m be a fraction with an even numerator, as* $\frac{2}{3}$. Then

$$y = b + c(x - a)^{\frac{2}{3}},$$

$$\frac{dy}{dx} = \frac{2c}{3(x-a)^{\frac{1}{3}}} \qquad \frac{d^2y}{dx^2} = -\frac{1}{3}\,\frac{2c}{3(x-a)^{\frac{4}{3}}};$$

$x = a$ gives

$$y = b \qquad \frac{dy}{dx} = \infty \qquad \frac{d^2y}{dx^2} = \infty.$$

c being first regarded as positive; if $x < a$, $\frac{dy}{dx}$ will be negative, and if $x > a$, it will be positive; hence at the point whose co-ordinates are $x = a$ and $y = b$, $\frac{dy}{dx}$ must change its sign from minus to plus, which change indicates a minimum ordinate, Art. (66).

If c be negative; the reverse will be the case, there will be a change of sign from plus to minus, and the ordinate will be a maximum.

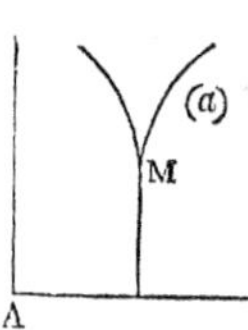

In the first case, the second differential coefficient for all values of x is negative, and the ordinate positive; the curve is therefore concave towards the axis of X, as represented in fig. (a).

In the second case, $\frac{d^2y}{dx^2}$ is always positive. For all positive values of y the curve will then be convex, and for all negative values of y concave, as in fig. (b). The tangent at the point M is in both cases perpendicular to the axis of X.

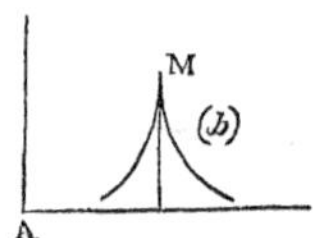

The point M is singular, and is called *a cusp.* It is a point at which the curve, when interrupted in its course in one direction, turns immediately into a contrary one.

5th. *Let m be a fraction with an even denominator, as $\frac{3}{2}$.*

Since the denominator of the fraction indicates that the square root is to be taken, the double sign $\pm$ must be placed before $(x-a)^{\frac{3}{2}}$, and we then have

$$y = b \pm c(x-a)^{\frac{3}{2}},$$

$$\frac{dy}{dx} = \pm \frac{3}{2}c(x-a)^{\frac{1}{2}}, \qquad \frac{d^2y}{dx^2} = \pm \frac{3c}{4(x-a)^{\frac{1}{2}}}.$$

Every value of $x < a$ gives y imaginary; $x = a$ gives $y = b$, and $x > a$ gives two values, one greater and the other less than b. There is then no point on the left of that one whose co-ordinates are $x = a$ and $y = b$; but on the right of this point the curve must extend indefinitely and consist of two branches.

$$x = a \qquad \text{gives} \qquad \frac{dy}{dx} = 0;$$

the tangent at M is then parallel to the axis of X.

Each value of $x > a$ gives two values for $\frac{d^2y}{dx^2}$, the one positive corresponding to the greater value of y, and the other negative; hence the upper branch is convex, and the lower concave, as in the figure, and the point M is a cusp.

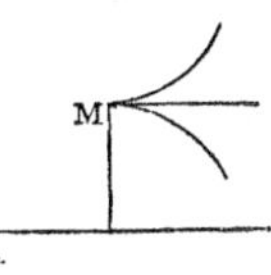

93. Let us now take the equation

$$(y - x^2)^2 = x^5,$$

from which we deduce

$$y = x^2 \pm x^{\frac{5}{2}}$$

$$\frac{dy}{dx} = 2x \pm \frac{5}{2}x^{\frac{3}{2}} \qquad \frac{d^2y}{dx^2} = 2 \pm \frac{5}{2}\,\frac{3}{2}x^{\frac{1}{2}}.$$

When $x = 0$, we have $y = 0$. If x be negative, y is imaginary. For every positive value of x, there are two real values of y, both of which are positive as long as $x^2 > x^{\frac{5}{2}}$ or $x < 1$; after which, one is positive and the other negative.

When $x = 0$, $\frac{dy}{dx} = 0$; also when

$$2 \pm \frac{5}{2}x^{\frac{1}{2}} = 0, \qquad \text{or} \quad x = \frac{16}{25},$$

hence the axis of X is tangent to the curve at the origin, and the tangent to the lower branch, at the point whose abscissa is $\frac{16}{25}$, is parallel to the axis of X.

The first value of $\frac{d^2y}{dx^2}$ belongs to the upper branch, and is always positive. The second value is also positive as long as $2 > \frac{5}{2}\cdot\frac{3}{2}\cdot x^{\frac{1}{2}}$, or $x < \frac{64}{225}$; after which it is negative.

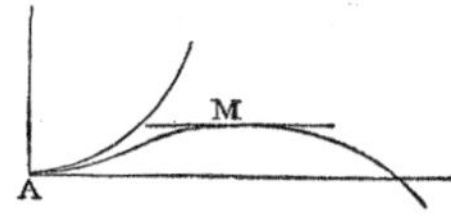

The origin is then a cusp, at which both branches lie on the same side of the common tangent, and is of the *second species*, those before discussed being of the *first species.* The point of the lower branch whose abscissa is $\frac{64}{225}$ is a point of inflexion.

94. By differentiating the equation

$$y = b \pm (x - a)\sqrt{x - c},$$

we derive

$$\frac{dy}{dx} = \pm\sqrt{x - c} \pm \frac{x - a}{2\sqrt{x - c}}.$$

For every value of $x < c$, y is imaginary.

For $x = c$, $y = b$, and $\frac{dy}{dx} = \infty$.

For every value of $x > c$, there are two real values of y.

For $x = a$, $y = b$, and $\frac{dy}{dx} = \pm\sqrt{a - c}$,

and at the corresponding point M there are two tangents, one making an angle, the tangent of which is $+\sqrt{a - c}$, and the other $-\sqrt{a - c}$. The point M is singular, and belongs to a class called *multiple points*, or points at which two or more branches of a curve intersect. If but two

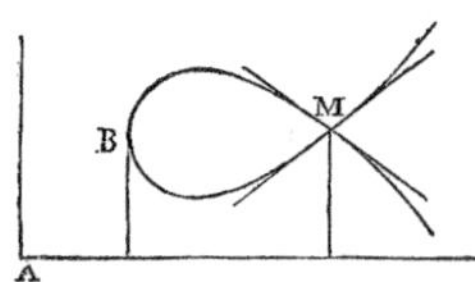

intersect, the point is a double multiple point; if three, a triple, and so on. Since there will be a separate tangent to each branch, at one of these points, it will be characterised by two or more values of the first differential coefficient, for the same values of the variables.

95. From the equation

$$ay^2 - x^3 + bx^2 = 0,$$

we derive

$$y = \pm\sqrt{\frac{x^2(x-b)}{a}} \qquad \frac{dy}{dx} = \pm\frac{3x - 2b}{2\sqrt{a(x-b)}}.$$

Since $x = 0$ gives $y = 0$, the origin A is a point of the curve. All negative values of x make y imaginary, as also all positive values less than b; hence A has no consecutive point. Such points, given by the equation of a curve, but having no consecutive points on either side, are singular, and are called *isolated* or *conjugate points.* At these points it is plain that no tangent can be drawn, and that therefore the *corresponding value of the first differential coefficient must be imaginary;* as in the above example, $x = 0$ gives

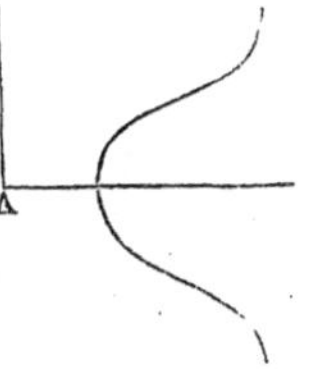

$$\frac{dy}{dx} = \frac{-b}{\sqrt{-ab}}.$$

96. We will close this branch of the subject by a discussion of the equation

$$ay^2 - x^3 + (b - c)x^2 + bcx = 0;$$

whence

$$y = \pm\sqrt{\frac{x(x-b)(x+c)}{a}}, \quad \frac{dy}{dx} = \pm\frac{3x^2 - 2x(b-c) - bc}{2\sqrt{ax(x-b)(x+c)}}.$$

Each of the values, $x = 0$, $x = b$, $x = -c$, gives $y = 0$.

Every negative value of $x > c$ gives y imaginary; while every such value less than c gives two equal values of y with contrary signs: Every positive value of $x < b$ gives y imaginary, and every such value greater than b, gives two equal values of y with contrary signs. The curve is then symmetrical with reference to the axis of X.

Each of the values, $x = 0$, $x = b$, $x = -c$, reduces $\frac{dy}{dx}$ to ∞; hence at the three corresponding points the tangent is perpendicular to the axis of X.

By solving the equation

$$3x^2 - 2x(b-c) - bc = 0,$$

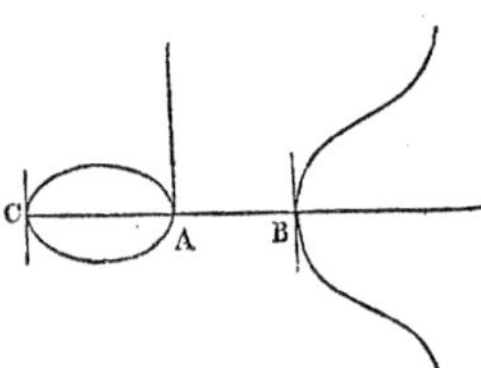

we shall find two real values for x, and thus determine the points at which the tangent is parallel to the axis of X. The curve may then be drawn as in the figure, in which AC $= -c$ and AB $= b$.

If $c = 0$, the equation becomes

$$ay^2 - x^3 + bx^2 = 0,$$

and the oval AC reduces to the conjugate point A, as in the preceding article.

If $b = 0$, the equation becomes

$$ay^2 - x^3 - cx^2 = 0,$$

and the curve takes the form indicated in figure (b), the origin being a double multiple point, since $\frac{dy}{dx}$ becomes equal to $\pm\sqrt{\frac{c}{a}}$.

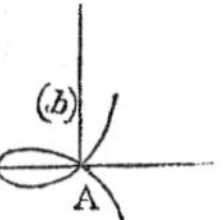

If b and c are both equal to 0, the equation becomes

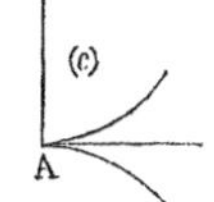

$$ay^2 - x^3 = 0; \quad \text{whence} \quad y = \pm\sqrt{\frac{x^3}{a}},$$

and the curve will be as in figure (c), the point A being a cusp of the first species.

OSCULATORY CURVES AND CURVATURE.

97. It is now proposed to examine the tendency which curves, with a common point, have to coincide with each other in the vicinity of this point; and also the use which may be made of this property of curves.

Let there be the three curves BB′, CC′, DD′, having the point M common; the co-ordinates of the first curve being represented by x and y, those of the second by x' and y', and those of the third by x'' and y''.

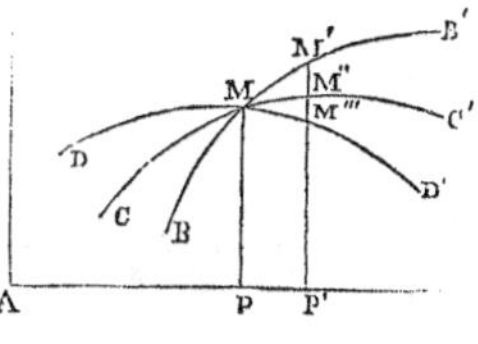

Since the point M is common, for this we have

$$\text{AP} = x = x' = x'' \qquad \text{PM} = y = y' = y''.$$

Suppose the abscissa AP to be increased by the variable h, we shall then have

$$\text{P}'\text{M}' = f(x+h) = y + \frac{dy}{dx}h + \frac{d^2y}{dx^2}\frac{h^2}{1.2} + \frac{d^3y}{dx^3}\frac{h^3}{1.2.3} + \&c.$$

$$P'M'' = f'(x' + h) = y' + \frac{dy'}{dx'}h + \frac{d^2y'}{dx'^2}\frac{h^2}{1.2} + \frac{d^3y'}{dx'^3}\frac{h^3}{1.2.3} + \&c$$

$$P'M''' = f''(x'' + h) = y'' + \frac{dy''}{dx''}h + \frac{d^2y''}{dx''^2}\frac{h^2}{1.2} + \frac{d^3y''}{dx''^3}\frac{h^3}{1.2.3} + \&c$$

in which

$$\frac{dy}{dx}, \qquad \frac{d^2y}{dx^2}, \qquad \frac{d^3y}{dx^3}, \ \&c.,$$

represent, what the first, second, &c., differential coefficients, obtained from the differential equations of the first curve, become by the substitution of the co-ordinates of the common point.

$\frac{dy'}{dx'}$, $\frac{d^2y'}{dx'^2}$, &c., are corresponding values for the second curve;

$\frac{dy''}{dx''}$, $\frac{d^2y''}{dx''^2}$, &c.......for the third.

By subtracting the second and third equations, each, member by member, from the first, and making

$$\frac{dy}{dx} - \frac{dy'}{dx'} = A, \qquad \frac{dy}{dx} - \frac{dy''}{dx''} = B,$$

$$\frac{d^2y}{dx^2} - \frac{d^2y'}{dx'^2} = A', \qquad \frac{d^2y}{dx^2} - \frac{d^2y''}{dx''^2} = B', \ \&c.,$$

we have

$$M'M'' = Ah + A'\frac{h^2}{1.2} + A''\frac{h^3}{1.2.3} + \&c.,$$

$$M'M''' = Bh + B'\frac{h^2}{1.2} + B''\frac{h^3}{1.2.3} + \&c.$$

Now, if h be made infinitely small, the points M′, M″, M‴, will become consecutive with the point M, and it is plain that the second curve will approach nearer to a coincidence with the first, than the third does, if M′M″ is numerically less than M′M‴, that is, if

$$Ah + A'\frac{h^2}{1.2} + A''\frac{h^3}{1.2.3} + \&c. < Bh + B'\frac{h^2}{1.2} + B''\frac{h^3}{1.2.3} + \&c.$$

This condition will necessarily be fulfilled if A is equal to 0, and B is not, as we shall have, after omitting the factor h,

$$A'\frac{h}{1.2} + A''\frac{h^2}{1.2.3} + \&c. < B + B'\frac{h}{1.2} + B''\frac{h^2}{1.2.3} + \&c.,$$

a true inequality when h is infinitely small, as then the whole of the first member will be less than the finite quantity B.

But $\qquad A = 0 \qquad$ gives $\qquad \dfrac{dy}{dx} = \dfrac{dy'}{dx'},$

that is, the first and second curves have a common tangent, or are tangent to each other at the common point.

If $A = 0$ and $B = 0$, the three curves have a common tangent, and in order that $M'M'' < M'M'''$, we must have

$$A'\frac{h^2}{1.2} + A''\frac{h^3}{1.2.3} + \&c. < B'\frac{h^2}{1.2} + B''\frac{h^3}{1.2.3} + \&c.,$$

which, it is proved as before, will necessarily be the case if $A' = 0$ and B′ is not. We have thus in addition the condition

$$\frac{d^2y}{dx^2} = \frac{d^2y'}{dx'^2}.$$

If $B' = 0$ or $\dfrac{d^2y}{dx^2} = \dfrac{d^2y''}{dx''^2}$ also, then $M'M'' < M'M'''$ if in addition to the other conditions we have

$$A'' = 0 \qquad \text{or} \qquad \frac{d^3y}{dx^3} = \frac{d^3y'}{dx'^3};$$

and in general the second curve will have a greater tendency than the third to coincide with the first, if the first and second have more equal successive differential coefficients of the ordinate at the common point, than the first and third.

Two curves which have a common point, and the first differential coefficients of the ordinate taken at this point equal to each other, are said to have a contact of the first order, or are simply tangent to each other.

If the first and second differential coefficients of the ordinate taken at this point are equal to each other respectively, the contact is of the second order.

And, in general, if the first m differential coefficients of the ordinate *taken at this point* are equal respectively, the contact is of the mth order.

To illustrate, take the two equations

$$y^2 = 4x \ldots\ldots(1), \qquad\qquad y = x + 1 \ldots\ldots(2).$$

By combining them we find a common point, the co-ordinates of which are

$$x'' = 1, \qquad\qquad y'' = 2.$$

By differentiation, we find from (1),

$$\frac{dy}{dx} = \frac{2}{y} \ldots\ldots(3); \qquad \text{whence} \qquad \frac{dy''}{dx''} = 1;$$

and from (2),

$$\frac{dy}{dx} = 1 \ldots\ldots(4); \qquad \text{"} \qquad \frac{dy''}{dx''} = 1.$$

Differentiating again, we have from (3),

$$\frac{d^2y}{dx^2} = -\frac{4}{y^3}; \qquad \text{whence} \qquad \frac{d^2y''}{dx''^2} = -\frac{1}{2};$$

and from (4),

$$\frac{d^2y}{dx^2} = 0 ; \qquad \text{whence} \qquad \frac{d^2y''}{dx''^2} = 0 ;$$

The two lines having a point in common, and the first differential coefficients of the ordinate equal at this point, have a contact of the first order. Since the second differential coefficients are not equal, the order of contact is no higher than the first.

98. The constants which enter into the equation of a curve determine its extent and position with respect to the co-ordinate axes. If then one curve be given completely, and another in kind only, by its general equation, the constants in this equation being arbitrary, we can evidently assign such values to them as shall cause the curve to fulfil as many conditions as its equation contains constants; that is, we may make the co-ordinates of one point of the second curve equal to those of a given point of the first; and, in addition, as many differential coefficients of the ordinate taken at this point, for the second curve, equal to the corresponding ones of the first, as there are constants to be disposed of, less one; thus giving to the second curve an order of contact at a given point of the first, *denoted by the number of constants less one.*

To ascertain the values which must be assigned to the arbitrary constants: Obtain first, the value of the ordinate from the equation of the second curve, (the abscissa being assumed equal to the abscissa of the given point,) and place it equal to the ordinate of the given point; or what amounts to the same thing, substitute the co-ordinates of the given point in the equation of the second curve; obtain then the first differential coefficients of the ordinate by differentiating the equation of each curve, substitute in these the co-ordinates of the given point, and place the results equal; do the same with the successive differential coefficients, until as many equations are formed as there are arbitrary constants. By the solution of these equations we can find those values of the constants which will cause the conditions to be fulfilled. These, substituted in the equation of the second curve, will give an equation which will represent the particular curve having the required order of contact.

The curve, which at a given point of a given curve has a higher order of contact than any other of the same kind, is called *an osculatrix*. Thus, *an osculatory circle is one which has a higher order of contact than any other circle.*

Since no more conditions can be assigned than there are constants; *the highest order of contact which can be given to a curve, is denoted by the number of constants less one, which enter its most general equation.*

Let these principles be applied:

1st. To find the equation of an osculatory right line.

Let the equation of the given curve be

$$y = f(x),$$

and the co-ordinates of the given point, x'' and y''. For this point, we have

$$y'' = f(x'').$$

The most general equation of the right line is

$$y = ax + b \ldots\ldots (1),$$

containing but two arbitrary constants. The first condition to be fulfilled is, that the value of y deduced from this equation, when $x = x''$, shall be equal to y'', that is

$$y'' = ax'' + b \ldots\ldots (2).$$

The first differential coefficient of the ordinate derived from the equation of the given curve is $\frac{dy}{dx}$, which for the given point becomes $\frac{dy''}{dx''}$. The first differential coefficient derived from equation (1) is $\frac{dy}{dx} = a$; hence the second condition is

$$\frac{dy''}{dx''} = a......(3).$$

By the solution of equations (2) and (3), we find

$$a = \frac{dy''}{dx''}, \qquad b = y'' - \frac{dy''}{dx''}x''.$$

These values in (1), give the equation

$$y = \frac{dy''}{dx''}x + y'' - \frac{dy''}{dx''}x'', \quad \text{or} \quad y - y'' = \frac{dy''}{dx''}(x - x'').$$

This, as it should be, is the same equation as that deduced in Art. (79).

2d. To find the equation of the osculatory circle at any point of the curve whose equation is $y = f(x)$.

Denote the given point, *or point of osculation*, by x'' and y''.

The most general equation of the circle is

$$(x - \alpha)^2 + (y - \beta)^2 = R^2......(1),$$

containing three arbitrary constants. A contact of the second order may therefore be given to the circle.

By differentiating the equation $y = f(x)$, and substituting x'' and y'' in the first and second differential coefficients, we obtain

$$\frac{dy''}{dx''} \quad \text{and} \quad \frac{d^2y''}{dx''^2}.$$

Differentiating equation (1) twice, we have

$$(x - \alpha)dx + (y - \beta)dy = 0; \quad \text{whence} \quad \frac{dy}{dx} = -\frac{x - \alpha}{y - \beta},$$

$$dx^2 + dy^2 + (y - \beta)d^2y = 0; \quad \text{whence} \quad \frac{d^2y}{dx^2} = -\frac{1 + \frac{dy^2}{dx^2}}{y - \beta}.$$

But the conditions that the circle be an osculatrix are, (x being assumed equal to x'',)

$$y = y'' \qquad \frac{dy}{dx} = \frac{dy''}{dx''} \qquad \frac{d^2y}{dx^2} = \frac{d^2y''}{dx''^2}.$$

We shall then have for the three equations of condition,

$$(x'' - \alpha)^2 + (y'' - \beta)^2 = R^2 \ldots\ldots(2),$$

$$\frac{dy''}{dx''} = -\frac{x'' - \alpha}{y'' - \beta} \ldots\ldots(3), \qquad \frac{d^2y''}{dx''^2} = -\frac{1 + \frac{dy''^2}{dx''^2}}{y'' - \beta} \ldots\ldots(4).$$

By the solution of these, we can find the values of R, α and β, which substituted in (1) will give the equation of the osculatory circle.

To illustrate, let us seek the equation of the circle osculatory to the parabola whose equation is

$$y^2 = 4x,$$

at the point whose co-ordinates are $x'' = 1, \quad y'' = 2.$

Differentiating the given equation twice, and substituting the co-ordinates 1 and 2, we find

$$\frac{dy}{dx} = \frac{2}{y}; \qquad \text{whence} \qquad \frac{dy''}{dx''} = 1;$$

$$\frac{d^2y}{dx^2} = -\frac{4}{y^3}; \qquad \text{"} \qquad \frac{d^2y''}{dx''^2} = -\frac{1}{2}.$$

These values, with the co-ordinates of the given point, placed in the equations of condition, give

$$(1 - \alpha)^2 + (2 - \beta)^2 = R^2$$

$$1 = -\frac{1 - \alpha}{2 - \beta} \qquad -\frac{1}{2} = -\frac{2}{2 - \beta};$$

whence

$$\alpha = 5 \qquad \beta = -2 \qquad R = \sqrt{32},$$

and the equation of the osculatory circle will then be

$$(x - 5)^2 + (y + 2)^2 = 32.$$

99. Since in the three equations of condition just considered, x'' and y'' may, in succession, be made to represent every point of the given curve, we may omit the dashes and write the equations thus

$$(x - \alpha)^2 + (y - \beta)^2 = R^2 \ldots\ldots\ldots\ldots (1),$$

$$x - \alpha = -\frac{dy}{dx}(y - \beta) \ldots\ldots\ldots\ldots (2),$$

$$y - \beta = -\frac{dx^2 + dy^2}{d^2y} \ldots\ldots\ldots\ldots (3);$$

in which it must be recollected, x and y are the co-ordinates of the point of osculation, α and β the co-ordinates of the centre of the osculatory circle, and R its radius.

Substituting in (1) the value of $x - \alpha$, and reducing, we obtain

$$R^2 = (y - \beta)^2 + \frac{dy^2}{dx^2}(y - \beta)^2 = (y - \beta)^2\left(\frac{dx^2 + dy^2}{dx^2}\right);$$

whence, by the substitution of the value of $y - \beta$,

$$R = \pm \frac{(dx^2 + dy^2)^{\frac{3}{2}}}{dx\, d^2y},$$

which is a general value for the radius of the osculatory circle.

If z denote the arc of the given curve, then

$$dz = \sqrt{dx^2 + dy^2} \text{ ....Art. (86)};$$

hence the above expression for R becomes

$$R = \pm \frac{dz^3}{dx\, d^2y}.$$

100. If φ denote the angle made by the radius of the osculatory circle drawn to the point of osculation, with a fixed line as OP, M and M′ two consecutive points, and MC and M′C the corresponding radii intersecting at C, then

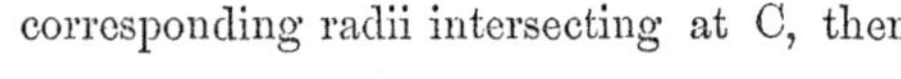

$$MC = R,\ MM' = dz,\ nn' = d\varphi \text{......Art. (91)}.$$

Since MCM′ may be regarded as a triangle right-angled at M, we have

$$MM' = MC \text{ tang } MCM',$$

and since MCM′ is infinitely small, the arc which measures it may be taken for its tangent; hence

$$dz = Rd\varphi, \qquad \text{and} \qquad R = \frac{dz}{d\varphi}.$$

101. The first value of R in article (99) has been deduced under the supposition that x is the independent variable. It is sometimes desirable to change this independent variable, during the discussion of expressions of this kind, and to regard y or some other variable quantity in the expression as the independent one. A more general expression for R may be obtained without the particular supposition referred to, if we recollect that $\frac{d^2y}{dx}$ has been deduced by the differentiation of $\frac{dy}{dx}$, regarding $d(dx) = 0$.

If we differentiate this expression on the supposition that both dy and dx are variable, we have

$$d\left(\frac{dy}{dx}\right) = \frac{dxd^2y - dyd^2x}{dx^2},$$

which must take the place of $\frac{d^2y}{dx}$, or for

$$dxd^2y \qquad \text{we must put} \qquad dxd^2y - dyd^2x.$$

The value of R thus becomes

$$R = \pm \frac{(dx^2 + dy^2)^{\frac{3}{2}}}{dxd^2y - dyd^2x} = \pm \frac{dz^3}{dxd^2y - dyd^2x} \ldots\ldots(1).$$

If in this, dx be regarded as constant, we shall have the value of R, as in article (99).

If dy be constant, or y regarded as the independent variable, then

$$R = \mp \frac{(dx^2 + dy^2)^{\frac{3}{2}}}{dyd^2x} = \mp \frac{dz^3}{dyd^2x}.$$

If z be regarded as the independent variable, dz will be constant, and $\quad d(dz^2) = 0; \quad$ whence

$$dxd^2x + dyd^2y = 0.$$

Adding the square of this, to the denominator of the value of R^2 taken from (1), we have

$$R^2 = \frac{dz^6}{[(d^2y)^2 + (d^2x)^2](dy^2 + dx^2)} = \frac{dz^4}{(d^2y)^2 + (d^2x)^2};$$

$$R = \pm \frac{dz^2}{\sqrt{(d^2y)^2 + (d^2x)^2}}.$$

102. Since the curve and osculatory circle at a given point have a tangent in common, they must also have the same normal; but the normal to the circle passes through its centre, the normal to the curve must then pass through this centre; or *the radius of the osculatory circle, drawn to the point of osculation, is normal to the curve.*

103. Let BB′ be any curve, and CC an arc of the osculatory circle. Then since $\frac{dy}{dx} = \frac{dy'}{dx'}$ and $\frac{d^2y}{dx^2} = \frac{d^2y'}{dx'^2}$, we shall have, Art. (97),

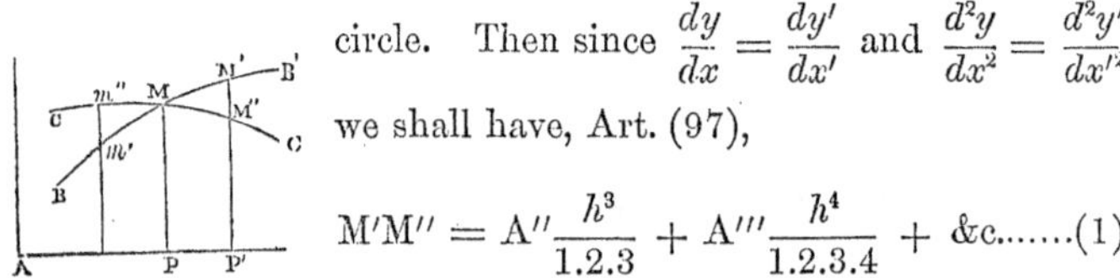

$$M'M'' = A''\frac{h^3}{1.2.3} + A'''\frac{h^4}{1.2.3.4} + \&c.......(1).$$

When h is infinitely small, the sign of M′M″ will depend upon that of the first term of the series, which will have the same sign as A″ when h is positive, and a contrary one when h is negative; that is, M′M″ and $m'm''$ have contrary signs. If then M″ is below the curve BB′, m'' will be above it, and the reverse; *and the circle* CC *must intersect the curve at* M.

It may be shown in the same way, that *any osculatrix of an even order intersects the curve; while one of an uneven order does not*. As, when the order of contact is even, the first term of (1) will contain h with an odd exponent, and will therefore change its sign when h becomes $-h$. This will not be the case when the power of h in the first term of (1) is denoted by an even number.

The osculatory circle, however, does not intersect at those points about which the curve is symmetrical with its normal. For, ordinates being drawn from the points of both, perpendicular to the common normal, if the ordinate of the curve on one side is greater than the corresponding ordinate of the circle, it will be so on the other side; as may be seen in the figure, in which, if $pn > po$, then $pn' > po'$; or if $pn < po$, then $pn' < po'$; hence, in this case, in the vicinity of the point M, the circle lies entirely within or entirely without the curve. In these cases it will be found that the order of contact of the circle is odd, and higher than the second, for unless $A'' = 0$, the circle must intersect, as shown by the preceding demonstration.

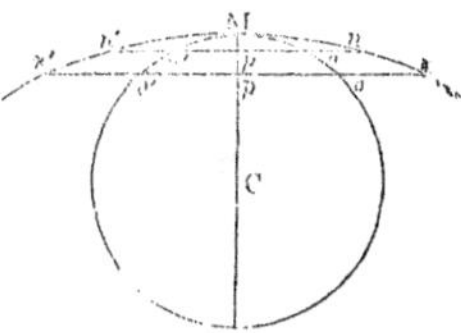

Since the osculatory circle has a more intimate contact with a curve at a given point than any other circle, it will necessarily separate those circles which are tangent without the curve from those which are tangent within.

104. *The curvature of a curve at a given point is its tendency to depart from its tangent at that point.* Thus, of the two curves AC and AB, having the common tangent AD, the former has a greater tendency to depart from the tangent, and has the greatest curvature.

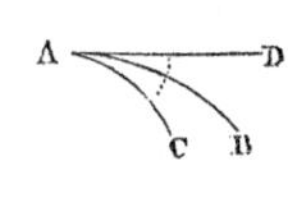

The curvature of the circumference of a circle, is evidently the same at all of its points, but of two different circumferences, that one curves the most which has the least radius; as in the figure,

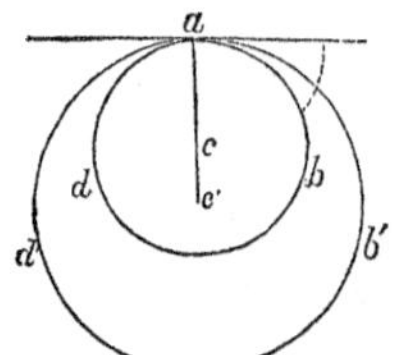

the tendency of *abd* to depart from the tangent is greater than that of *ab'd'*, and this tendency plainly increases as the radius decreases, and the reverse; that is, *the curvature in two different circles varies inversely as their radii.*

This being the case, the expression $\frac{1}{R}$ may be taken as the measure of the curvature of a circle whose radius is R.

Since the contact of the osculatory circle with a curve is so intimate, its curvature may be taken for the curvature of the curve at the point of osculation; and the two in the immediate vicinity of this point, may be regarded as one and the same curve; hence, to compare the curvatures at different points of a curve, we have only to compare the curvatures of the osculatory circles drawn at these points. Thus in the curve MM',

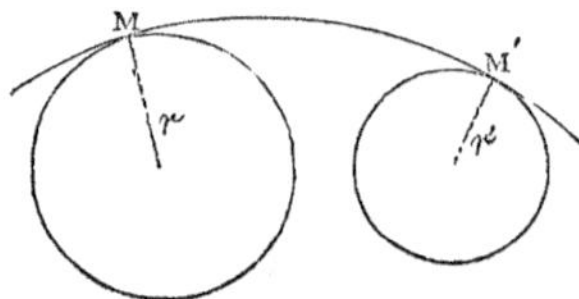

$$\textit{curvature at } M : \textit{curvature at } M' :: \frac{1}{r} : \frac{1}{r'}.$$

105. The radius of the osculatory circle at a given point of a curve is called *the Radius of Curvature*, at that point. The general value of this radius is given in article (99), and it may be found for any particular curve, by differentiating the equation of the curve, and substituting the derived values of dy and d^2y in the formula,

$$R = \pm \frac{(dx^2 + dy^2)^{\frac{3}{2}}}{dxd^2y}.$$

If the value at any particular point of the curve be required for x and y in the value just deduced, substitute the co-ordinates of the particular point.

As only the absolute length of the radius of curvature is required in determining the curvature of curves, we may use either the plus or minus sign of the formula. It is best, in general, to use that which, taken with the sign resulting from the expression, will make R essentially positive.

Let it now be required to find the general expression for the radius of curvature of Conic Sections.

Their equation is

$$y^2 = mx + nx^2; \qquad \text{whence} \qquad dy = \frac{(m + 2nx)dx}{2y},$$

$$dx^2 + dy^2 = \frac{[4y^2 + (m + 2nx)^2]dx^2}{4y^2},$$

$$d^2y = \frac{2nydx^2 - (m + 2nx)dxdy}{2y^2} = \frac{[4ny^2 - (m + 2nx)^2]dx^2}{4y^3}.$$

These values substituted in the formula, give

$$R = \frac{[4(mx + nx^2) + (m + 2nx)^2]^{\frac{3}{2}}}{2m^2},$$

and this, after dividing both terms of the fraction by 8, may be put under the form

$$R = \frac{\left(\sqrt{mx + nx^2 + \frac{1}{4}(m + 2nx)^2}\right)^3}{\frac{m^2}{4}} \text{............}(1);$$

the numerator of which is the cube of the normal, Art. (82): Hence the radius of curvature at any point of a conic section, is the *cube of the normal divided by the square of half the para-*

meter, and the radii at different points are to each other as the cubes of the corresponding normals.

If in (1) we make $x = 0$, we have, at the principal vertex,

$$R = \frac{m}{2} = \textit{one half the parameter},$$

which for the ellipse and hyperbola is $\frac{B^2}{A}$.

The radius of curvature at the vertex of the conjugate axis of the ellipse is obtained by substituting in (1),

$$m = \frac{2B^2}{A}, \qquad n = -\frac{B^2}{A^2}, \qquad \text{and} \qquad x = A.$$

The result is

$$R = \frac{A^2}{B} = \textit{one half the parameter of the conjugate axis.}$$

It may be readily shown that $\frac{m}{2}$ is *the least value* which R admits of; therefore the curvature at the principal vertex of a conic section is greater than at any other point. Likewise, $\frac{A^2}{B}$ is *the greatest value* of R in the ellipse; hence the curvature of the ellipse *is least*, at the vertex of the conjugate axis. The curvature of the other two curves diminishes as we recede from the vertex.

For the parabola $n = 0$; we then have

$$R = \frac{(m^2 + 4mx)^{\frac{3}{2}}}{2m^2}.$$

EVOLUTES.

106. If at the different points of a given curve osculatory circles be drawn, and a second curve traced through their centres, the latter is called *the Evolute* of the former, which is *the Involute*. Thus CC''' is *the evolute* of the involute MM'''. Points of the evolute may always be constructed by drawing normals at the different points of the involute, and on each of these normals laying off the corresponding value of R, deduced as in article (105).

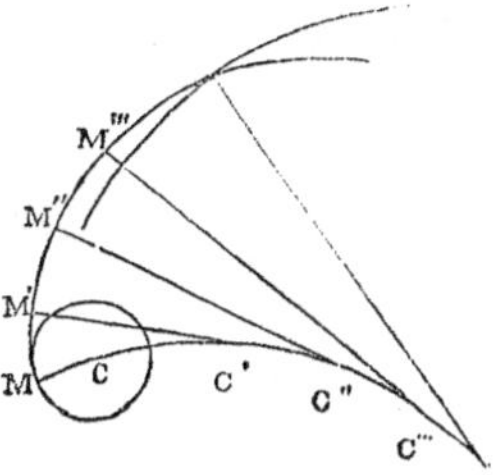

107. If α and β, the co-ordinates of the centre of the osculatory circle, be regarded as variables, they will determine all the points of the evolute; but α, β, and R, are functions of x and y, the co-ordinates of the points of osculation; and the relation between these five variables is expressed by the three equations of Art. (99), which may be written thus,

$$(x-\alpha)^2 + (y-\beta)^2 = R^2 \ldots\ldots\ldots\ldots (1),$$

$$(x-\alpha)dx + (y-\beta)dy = 0 \ldots\ldots\ldots (2),$$

$$(y-\beta)d^2y + dy^2 + dx^2 = 0 \ldots\ldots\ldots (3).$$

If we differentiate (1) and (2), regarding all the quantities, except dx, as variables, we obtain

$$(x-\alpha)dx + (y-\beta)dy - (x-\alpha)d\alpha - (y-\beta)d\beta = RdR,$$

$$dx^2 + dy^2 + (y-\beta)d^2y - dxd\alpha - dyd\beta = 0,$$

and these, by means of equations (2) and (3), are reduced to

$$-(x-\alpha)d\alpha - (y-\beta)d\beta = \mathrm{R}d\mathrm{R}\ldots\ldots(4),$$

$$-dxd\alpha - dyd\beta = 0.\ldots\ldots\ldots\ldots(5).$$

Equation (5) gives

$$-\frac{dx}{dy} = \frac{d\beta}{d\alpha}\ldots\ldots\ldots\ldots(6).$$

$-\frac{dx}{dy}$ is the tangent of the angle which a normal to the involute at the point (x, y) makes with the axis of X, Art. (81), and $\frac{d\beta}{d\alpha}$ is the tangent of the angle which a tangent to the evolute at the point (α, β) makes with the same axis; hence these angles are equal. But the normal at the point (x, y) passes through the point (α, β), Art. (102); therefore the normal and tangent form one and the same line; that is, *the radius of curvature is normal to the involute, and tangent to the evolute.*

The evolute may therefore be constructed, by drawing a curve tangent to the normals at the different points of the involute.

From what precedes, it is plain that the evolute may be regarded as formed by the intersections of the consecutive normals to the involute, and that the point of intersection of any two consecutive normals may be taken as the centre of the osculatory circle, which passes through the two consecutive points of the involute at which the normals are drawn.

108. Equation (6) of the preceding article, combined with (2), gives

$$x - \alpha = \frac{d\alpha}{d\beta}(y - \beta).$$

Substituting this value in (1), we have, after reduction,

$$(y - \beta)^2 \frac{(d\alpha^2 + d\beta^2)}{d\beta^2} = \mathrm{R}^2 \ldots\ldots\ldots\ldots (7).$$

Substituting the same value in (4), reducing and squaring both members, we obtain

$$(y - \beta)^2 \frac{(d\alpha^2 + d\beta^2)^3}{d\beta^2} = \mathrm{R}^2 d\mathrm{R}^2.$$

Dividing this by (7), member by member, and taking the root,

$$\sqrt{d\alpha^2 + d\beta^2} = d\mathrm{R}.$$

But if z represent the arc of the evolute, we have

$$dz = \sqrt{d\alpha^2 + d\beta^2} \ldots\ldots\ldots\ldots \text{Art. } (86);$$

hence

$$d\mathrm{R} = dz, \qquad d\mathrm{R} - dz = 0, \qquad d(\mathrm{R} - z) = 0;$$

whence $\mathrm{R} - z$ must be a constant, Art. (14), or

$$\mathrm{R} = z + c.$$

109. If any two radii of curvature be drawn, as one at M and the other at M′; the first being denoted by R, the second by R′, and the corresponding arcs BC and BC′ by z and z', we have

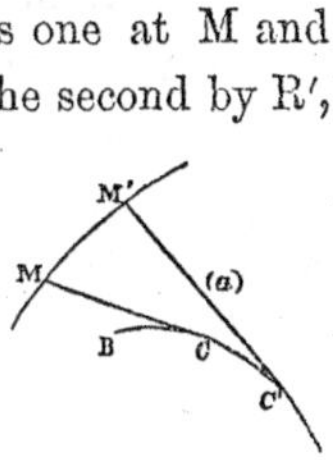

$$\mathrm{R} = z + c \qquad \mathrm{R}' = z' + c;$$

whence

$$\mathrm{R}' - \mathrm{R} = z' - z;$$

that is, *the difference between any two radii of curvature is equal to the arc of the evolute intercepted between them.*

If in the equation $R = z + c$, we make $z = 0$, and denote by r, the corresponding value of R, we shall have

$$r = 0 + c = c;$$

that is, the constant c is always equal to the radius of curvature which passes through the point of the evolute, from which its arc is estimated.

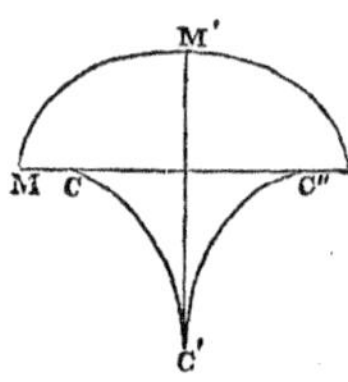

If we estimate the evolute of the ellipse from the point C, we have

$$c = MC = \frac{B^2}{A} \text{............Art. (105).}$$

hence

$$R = z + \frac{B^2}{A}.$$

Also, since $$M'C' = \frac{A^2}{B},$$

$$M'C' - MC = \frac{A^2}{B} - \frac{B^2}{A} = CC'.$$

If the evolute and one point of the involute be given, and a thread be wound upon the evolute and drawn tight, passing through the given point M, fig. (a); when the thread is unwound or *evolved*, the point of a pencil first placed at M, will describe the involute; for, by the nature of the operation, CC′ is always equal to M′C′ − MC.

110. The equation of the evolute of any curve may be found

thus: Differentiate the equation of the involute twice; deduce the expressions for dy and d^2y, and substitute in the equations

$$\left.\begin{aligned} y - \beta &= -\frac{dx^2 + dy^2}{d^2y} \text{..............}(1) \\ x - \alpha &= -\frac{dy}{dx}(y - \beta) \text{............}(2) \end{aligned}\right\} \text{Art. (99)};$$

combine the results, which will contain the four variables α, β, x, and y, with the equation of the involute, and eliminate x and y; the final equation will contain only α, β, and constants, and will therefore be the required equation.

As an example; let it be required to find the equation of the evolute of the common parabola.

The equation of the involute is

$$y^2 = 2px; \qquad \text{whence} \qquad \frac{dy}{dx} = \frac{p}{y};$$

$$dy^2 = \frac{p^2dx^2}{y^2}, \qquad\qquad d^2y = -\frac{p^2dx^2}{y^3}.$$

Substituting these values in (1) and (2), and reducing, we have

$$y - \beta = \frac{y^3}{p^2} + y; \qquad \text{whence} \qquad -\beta = \frac{y^3}{p^2} \text{......}(3);$$

$$x - \alpha = -\frac{y^2}{p} - p \text{............}(4);$$

and putting for y, in (3) and (4), its value $\sqrt{2px} = (2p)^{\frac{1}{2}}x^{\frac{1}{2}}$, we have

$$-\beta = \frac{2^{\frac{3}{2}}x^{\frac{3}{2}}}{p^{\frac{1}{2}}}, \qquad x - \alpha = -2x - p.$$

The value of $x = \frac{1}{3}(\alpha - p)$ taken from the last equation, and substituted in the preceding, gives

$$\beta^2 = \frac{8}{27p}(\alpha - p)^3,$$

which is the required equation.

If we make $\beta = 0$, we have $\alpha = p$, and laying off $AC = p$, C will be the point at which the evolute meets the axis of X. If we transfer the origin of co-ordinates to this point, we have

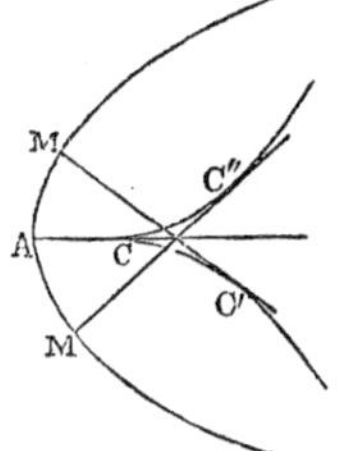

$$\alpha' = \alpha - p, \qquad \beta' = \beta;$$

hence

$$\beta'^2 = \frac{8}{27p}\alpha'^3.$$

Since every value of α' gives two values of β', equal with contrary signs, the curve is symmetrical with the axis of X. If α' be negative, β' is imaginary, and the curve does not extend to the left of C. The branch CC' belongs to AM, and CC'' to AM'.

TRANSCENDENTAL CURVES.

111. The most general division of curves is into the classes, *Algebraic* and *Transcendental*.

When the relation between the ordinate and abscissa of a curve can be expressed entirely in algebraic terms [see Art. (5)], it belongs to the first class; and when such relation can not be expressed without the aid of transcendental quantities, it belongs to the second class.

112. One of the most important of the latter class is

THE LOGARITHMIC CURVE,

so named, because it may always be referred to a set of co-ordinate axes, such that one co-ordinate will be the logarithm of the other. Its equation is usually written

$$y = \log x,$$

or, if a be the base of the system of logarithms,

$$x = a^y.$$

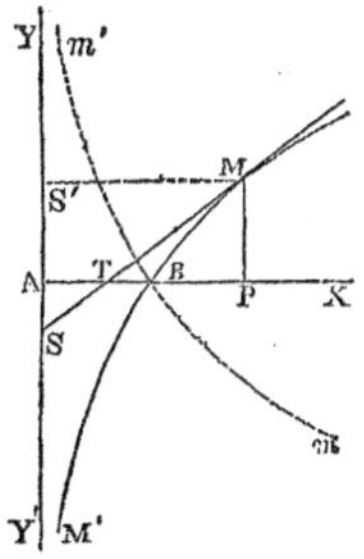

The curve is given when a is known, and may be constructed by laying off on the axis of X the different numbers, and on the corresponding perpendiculars, the logarithms of these numbers: Or it may be constructed from the equation $x = a^y$, by making $y = \frac{1}{2}$, $\frac{3}{2}$, $\frac{1}{4}$, &c.; whence the corresponding values of x are

$$x = \sqrt{a}, \qquad x = a\sqrt{a}, \qquad x = \sqrt[4]{a}, \text{ \&c.}$$

When $y = 0$, $x = 1$. This being the case for all systems of logarithms, shows that all logarithmic curves, when referred to the same axes, cut the axis of X, or *axis of numbers*, at a distance from the origin equal to unity.

If $a > 1$, and $x > 1$, y is positive and increases as x increases; if $x < 1$, y is negative and increases numerically as x decreases until $x = 0$, when $y = -\infty$. If x be negative, there will be no corresponding value of y. The curve will then be of the form indicated by the full line in the figure.

If $a < 1$, the reverse will be the case, and the curve will be represented by the dotted line.

113. If now we differentiate the equation $y = \log x$, M being the modulus, we deduce

$$\frac{dy}{dx} = \frac{M}{x}, \qquad \frac{d^2y}{dx^2} = -\frac{M}{x^2}.$$

When $\quad x = 0 \quad \frac{dy}{dx} = \frac{M}{0} = \infty;$

hence the tangent at the corresponding point is the axis of Y; and since for $x = 0$, $y = -\infty$, this tangent is an asymptote.

When $\quad x = \infty, \quad \frac{dy}{dx} = \frac{M}{\infty} = 0.$

But $x = \infty$ gives $y = \infty$; hence there is no tangent parallel to the axis of X, at a finite distance from it.

The value for the subtangent on the axis of X is

$$PT = y\frac{dx}{dy} = \log x \frac{x}{M}.$$

If the subtangent be taken on the axis of Y, we have

$$SS' = x\frac{dy}{dx} = M.$$

that is, *the subtangent on the axis of logarithms is constant,* and equal to the modulus of the system in which the logarithms are taken.

If $M = 1$, $SS' = 1 = AB$.

Since, when $a > 1$, $\frac{d^2y}{dx^2}$ is negative for all values of x, the part BM is concave towards the axis of X, and BM′ convex.

When $a < 1$, M is negative, $\frac{d^2y}{dx^2}$ will be positive, the part Bm' convex and Bm concave.

114. Another remarkable transcendental curve is,

THE CYCLOID,

which is generated by a point in the circumference of a circle, when the circle is rolled in the same plane, along a given straight line.

Let AB be the given line, and suppose the circle to have been placed upon it, so that the generating point was at A, and then to have been rolled to the position RME.

The generating point now at M, has generated the arc AM.

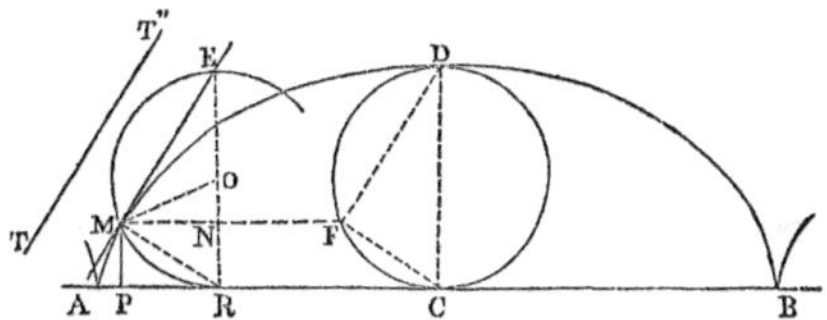

Take the origin of co-ordinates at A, and let $AP = x$, $PM = y$ and RE, the diameter of the generating circle $= 2r$; then

$$AP = AR - PR \text{............} (1).$$

But since every point of the circumference from M to R, as the circle was rolled, came in contact with AR, we have

$$AR = \text{arc } MR = \text{ver-sin}^{-1} RN = \text{ver-sin}^{-1} y.$$

Also,

$$\text{PR} = \text{MN} = \sqrt{\text{RN} \times \text{NE}} = \sqrt{y(2r - y)} = \sqrt{2ry - y^2}.$$

Substituting the values of AP, AR and PR in (1), we have

$$x = \text{ver-sin}^{-1} y - \sqrt{2ry - y^2} \ldots\ldots\ldots\ldots(2),$$

which is the equation of the Cycloid.

After the circle has been rolled over once, every point of the circumference will have been in contact with AB, and the generating point will have arrived at B; we have then

$$\text{AB} = \textit{circumference of generating circle} = 2\pi r.$$

The given line is called the base of the Cycloid, and the line CD $= 2r$ perpendicular to AB at its middle point, is the axis.

If the rolling of the circle be continued beyond the point B, an infinite number of arcs, each equal to ADB, will be generated.

Every negative value of y in equation (2) makes x imaginary; hence there is no point of the curve below the axis of X.

$$y = 2r, \qquad \text{gives} \qquad x = \text{ver-sin}^{-1} 2r = \pi r = \text{AC}.$$

Every value of $y > 2r$ makes x imaginary; hence the greatest ordinate of the curve is equal to the diameter of the generating circle.

By differentiating (2) we have, Art. (42),

$$dx = \frac{rdy}{\sqrt{2ry - y^2}} - \frac{rdy - ydy}{\sqrt{2ry - y^2}};$$

or reducing

$$dx = \frac{ydy}{\sqrt{2ry - y^2}} \ldots\ldots\ldots\ldots\ldots\ldots(3),$$

which is the differential equation of the Cycloid.

115. Substituting the preceding value of dx in the formulas of article (82), and reducing, we have

$$\textit{Subtangent}, \text{ PT} = \frac{y^2}{\sqrt{2ry - y^2}}.$$

$$\textit{Tangent}, \text{ MT} = \frac{y\sqrt{2ry}}{\sqrt{2ry - y^2}}.$$

$$\textit{Subnormal}, \text{ PR} = \sqrt{2ry - y^2}.$$

$$\textit{Normal}, \text{ MR} = \sqrt{2ry}.$$

Since the subnormal $\text{PR} = \sqrt{2ry - y^2} = \text{MN}$, the diameter ER and normal MR intersect the base at the same point. Hence, to construct the normal at a given point, join it with the point at which the corresponding position of the generating circle is tangent to the base: Or, upon the greatest ordinate CD as a diameter, describe a circle, and, through the given point M, draw a line parallel to the base, from the point F in which it cuts the circle, draw the two chords CF and DF to the extremities of the diameter; a line through the given point parallel to CF will be the normal, and one parallel to DF the tangent.

If it be required to draw a tangent parallel to a given line as T′T″; draw the chord DF parallel to the given line, from F draw FM parallel to the base; the point M is the point of contact, through which draw a line parallel to T′T″.

116. From equation (3), article (114), we have

$$\frac{dy}{dx} = \frac{\sqrt{2ry - y^2}}{y} = \sqrt{\frac{2r}{y} - 1} \dots\dots\dots\dots(1),$$

which becomes 0 when $y = 2r$, and ∞ when $y = 0$; hence at the extremity of the greatest ordinate, the tangent is parallel to the base; and at the points A, B, &c., where the curve meets the base, it is perpendicular.

If we square both members of equation (1), we have

$$\frac{dy^2}{dx^2} = \frac{2r}{y} - 1.$$

Differentiating both members of this, we have

$$\frac{2dyd^2y}{dx^2} = -\frac{2rdy}{y^2}, \quad \text{or} \quad \frac{d^2y}{dx^2} = -\frac{r}{y^2}.$$

This second differential coefficient being negative for all values of y, the curve is concave towards the axis of X, Art. (83).

117. Substituting the values of dy and d^2y in the expression

$$R = -\frac{(dx^2 + dy^2)^{\frac{3}{2}}}{dxd^2y},$$

we obtain

$$R = \frac{\left(\frac{2rydx^2}{y^2}\right)^{\frac{3}{2}}}{\frac{rdx^3}{y^2}} = 2^{\frac{3}{2}}r^{\frac{1}{2}}y^{\frac{1}{2}} = 2\sqrt{2ry};$$

or since $\sqrt{2ry}$ is the expression for the normal, Art. (115), *the Radius of Curvature is equal to twice the normal at the point of osculation.*

If $\quad y = 0, \quad R = 0; \qquad$ and if $\quad y = 2r, \quad R = 4r;$

hence the radius of curvature at A, (see figure in next article) is equal to 0; and at D is $4r$; therefore, Art. (109), the arc AA' $= 4r$.

118. To obtain the equation of the evolute let us substitute the values of dy and d^2y in equations (1) and (2) of article (110).

After reduction, we find

$$y - \beta = 2y, \qquad x - \alpha = -2\sqrt{2ry - y^2};$$

whence

$$y = -\beta, \qquad x = \alpha - 2\sqrt{-2r\beta - \beta^2}.$$

These values, in the equation of the involute, Art. (114), give

$$\alpha = \text{ver-sin}^{-1} - \beta + \sqrt{-2r\beta - \beta^2} \ldots\ldots (1),$$

for the required equation.

If we produce DC to A' making CA' $=$ DC, and then transfer the origin to A', the new axes being A'X' and A'D, and the new co-ordinates α' and β', we shall have for any point, as M',

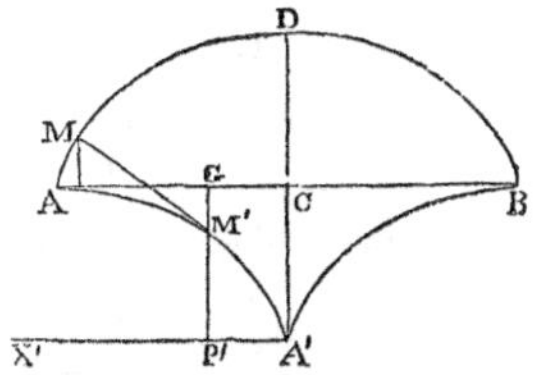

$$AG = \alpha, \qquad GM' = -\beta,$$

$$A'P' = \alpha', \qquad P'M' = \beta'.$$

Since AC $= \pi r$, and CG $=$ A'P',

$$\alpha = \pi r - \alpha';$$

and since GP' $= 2r$,

$$GM' = 2r - \beta', \qquad \text{or} \qquad -\beta = 2r - \beta'.$$

Substituting these values in (1), we have

$$\pi r - \alpha' = \text{ver-sin}^{-1}(2r - \beta') + \sqrt{2r\beta' - \beta'^2}$$

whence

$$\alpha' = \pi r - \text{ver-sin}^{-1}(2r - \beta') - \sqrt{2r\beta' - \beta'^2};$$

But $\pi r - \text{ver-sin}^{-1}(2r - \beta') = \text{ver-sin}^{-1}\beta'$; hence the last equation becomes

$$\alpha' = \text{ver-sin}^{-1}\beta' - \sqrt{2r\beta' - \beta'^2},$$

which is the equation of the evolute referred to the new axes, and is of the same form and contains the same constants as the equation of the involute, therefore the two curves are equal.

Since arc $AA' = 4r$, its equal $AD = 4r$, and $ADB = 4.2r$; that is, *equal to four times the diameter of the generating circle.*

THE SPIRALS.

119. If a right line be revolved uniformly, in the same plane, about one of its points; a second point of the line continually approaching, or receding from the fixed point, in accordance with some prescribed law, will generate a curve called *a spiral.*

The fixed point is called the *pole* or eye of the spiral. The portion of the spiral generated while the line makes one revolution, is called *a spire;* and since there is no limit to the number of revolutions, the number of spires is infinite, and any straight line drawn through the pole of the spiral will intersect it in an infinite number of points.

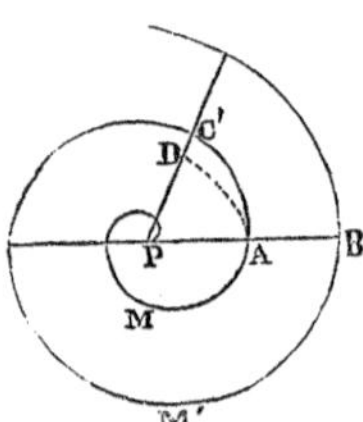

The system of polar co-ordinates being used to determine the different points of a spiral, its equation will, in general, be of the form

$$u = f(t),$$

in which u denotes the radius vector, and t the variable angle.

120. Before discussing the particular spirals, it will be necessary to determine general expressions for the subtangent, &c., and the differentials of the arc and area, in terms of polar co-ordinates.

The subtangent, in such case, is the part of the perpendicular to the radius vector of the point of contact, intercepted between the pole and the point where the tangent meets this perpendicular. Thus, if A be the pole, and MT the tangent, AT perpendicular to AM is the subtangent. To find the expression for it; let the arc t receive the increment PP′, (AP being $= 1$); describe MC with the radius AM $= u$; draw the chords MC and MM′, and the line AT′ parallel to MC, and produce MM′ to T′. From the similar triangles MM′C and M′AT′, we have

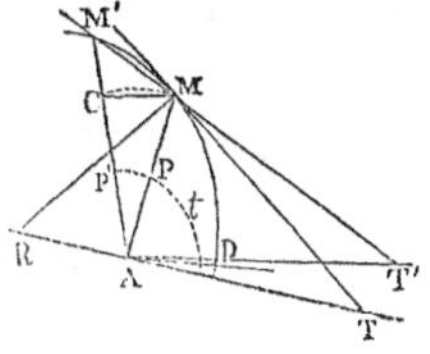

$$\text{M'C} : \text{MC} :: \text{M'A} : \text{T'A}; \qquad \text{T'A} = \frac{\text{MC} \times \text{M'A}}{\text{M'C}} \ldots\ldots(1).$$

Also from the similar sectors APP′ and AMC,

$$1 : \text{PP'} :: \text{AM} : \text{arc MC}; \qquad \text{arc MC} = \text{AM} \times \text{PP'}.$$

Now suppose the increment PP′ $= dt$, then M′C $= du$, Art. (91), M′ becomes consecutive with M, the secant M′T′ coincides with the tangent MT, T′A = AT, AM′ = AM $= u$ and chord MC = arc MC $= udt$.

Making these substitutions in (1), we have

$$\text{AT} = subtangent = \frac{u^2dt}{du} \ldots\ldots\ldots\ldots(2).$$

From this we deduce

$$\frac{\text{AT}}{u} = \frac{\text{AT}}{\text{AM}} = \frac{udt}{du} = \text{tang AMT}.$$

$$\textit{The tangent } \text{MT} = \sqrt{\overline{\text{AM}}^2 + \overline{\text{AT}}^2} = u\sqrt{1 + u^2\frac{dt^2}{du^2}}.$$

The similar triangles AMT and AMR, give

$$\text{AT} : u :: u : \text{AR}; \qquad \text{AR} = \frac{u^2}{\text{AT}} = \frac{du}{dt} = \textit{subnormal.}$$

When M′ is consecutive with M, MM′C may be regarded as a triangle, right-angled at C; hence

$$\text{MM}' = \sqrt{\overline{\text{M}'\text{C}}^2 + \overline{\text{MC}}^2}.$$

But MM′ is the differential of the arc; therefore

$$dz = \sqrt{du^2 + u^2dt^2}.$$

If ADM be any segment, AMM′ will be its increment when t is increased by dt. Calling the segment s, AMM′ will then be ds, and may be measured by the sector AMC. But the area of the sector

$$\text{AMC} = \frac{1}{2}\text{AM} \times \text{arc MC} = \frac{u^2dt}{2};$$

hence

$$ds = \frac{u^2dt}{2}.$$

It should be observed, that all of these expressions may be found precisely as in the corresponding cases in rectilinear co-ordinates, but it is better to avail ourselves of the more simple process indicated in the general remark, Art. (91).

121. An equation, from which the particular equations of most of the spirals may be deduced by assigning particular values to a and n, is

$$u = at^n$$

If n be positive, $t = 0$ will give $u = 0$,

and the spirals represented by the equation have their origin at the pole.

If n be negative, $t = 0$ will give $u = \infty$,

and the spirals have their origin at an infinite distance, continually approach the pole, and u becomes equal to 0 only when $t = \infty$.

122. Let $n = 1$, then $u = at$,

and if u' and t', u'' and t'', represent the co-ordinates of any two points of the spiral, we shall have

$$u' = at', \qquad u'' = at'';$$

whence

$$u' : u'' :: t' : t'',$$

or the law in accordance with which the generating point must move is, *that the radius vectors shall be proportional to the corresponding angles.*

The curve thus generated is the *Spiral of Archimedes.*

If we take foı the unit of distance, the length of the radius vector after one revolution; then $u = 1$, $t = 2\pi$, and the equation gives

$$1 = a.2\pi, \qquad a = \frac{1}{2\pi},$$

and the primitive equation becomes

$$u = \frac{t}{2\pi}; \qquad \text{whence} \qquad du = \frac{dt}{2\pi}.$$

This spiral may be constructed by dividing a circumference into any number of equal parts, as 8, and the radius AB into the same number of equal parts. On the radius AC lay off one of these parts; on AD two, AE three, &c.; on AB eight, then again on AC nine, &c. The distances thus laid off will be proportional to the angles BAC, BAD, &c., and the curve through their extremities the required spiral.

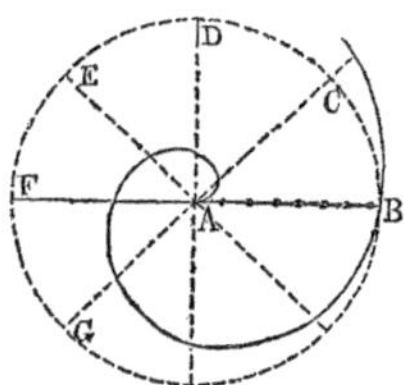

Substituting the values of u and du in equation (2), Art. (120), we have

$$\text{AT} = \textit{subtangent} = \frac{t^2}{2\pi}.$$

If $t = 2\pi$, that is, if the tangent be drawn at the extremity of the arc generated in one revolution, we have

$$\text{AT} = 2\pi = \textit{circumference of measuring circle.}$$

If $t = m.2\pi$, or the tangent be drawn at the extremity of the arc generated in m revolutions,

$$\text{AT} = m^2.2\pi = m.2m\pi\ ;$$

that is, equal to m *times the circumference described with the radius vector of the point of contact.*

For the subnormal we find

$$\text{AR} = \frac{du}{dt} = \frac{1}{2\pi}.$$

123. If $n = \frac{1}{2}$, the general equation becomes

$$u = at^{\frac{1}{2}}, \qquad \text{or} \qquad u^2 = a^2t.$$

This equation being of the same form as that of the parabola, the curve given by it is called the *Parabolic Spiral.*

It may be constructed by first constructing the parabola whose equation is $y^2 = a^2x$, and then laying off from P to B, C, D, &c., along the circumference, any assumed abscissas, and from A to M, M′, &c., the corresponding ordinates; the points M, M, &c., will be points of the spiral, since for each we have

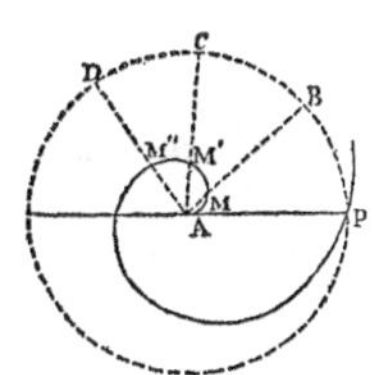

$$y^2 = a^2x, \qquad \text{or} \qquad u^2 = a^2t.$$

The subtangent at any point is $\text{AT} = \dfrac{2u^3}{a^2}.$

124. If $n = -1$, $u = at^n$ becomes

$$u = at^{-1} = \frac{a}{t}, \qquad \text{or} \qquad ut = a,$$

and the spiral thus given is called the *Hyperbolic Spiral.*

If u' and t', u'' and t'', be the co-ordinates of any two points of the spiral, we have $u' = \frac{a}{t'}$, and $u'' = \frac{a}{t''}$; whence

$$u' : u'' :: \frac{1}{t'} : \frac{1}{t''},$$

or *the radius vectors are inversely proportional to the angles.*

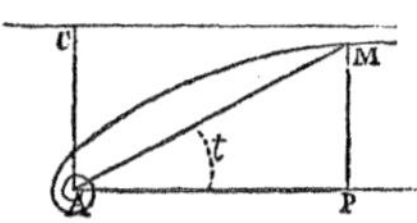

If M be any point of the spiral,

$$\text{AM} = u, \qquad \text{MAP} = t.$$

The right-angled triangle MAP, gives

$$u = \frac{\text{MP}}{\sin t}.$$

Substituting this value of u in the equation $ut = a$, we find

$$\text{MP} = a\frac{\sin t}{t}.$$

As t is diminished, this value approaches nearer to a, and since $\left(\frac{\sin t}{t}\right)_{t=0} = 1$; when $t = 0$, we have MP $= a$.

If then at a distance AC $= a$, a line be drawn parallel to AP, it will continually approach the curve and touch it at an infinite distance.

$$\textit{The subtangent } \text{AT} = \frac{u^2dt}{du} = -a.$$

It is then constant and equal to AC. Also,

$$\frac{udt}{du} = \text{tang AMT} = -t;$$

that is, *the tangent of the angle made by the tangent and radius vector is equal to the arc which measures the angle made by the radius vector and fixed line.*

We may apply these properties to the construction of the curve by points, thus: With A as a centre and radius $= a$, describe a circle; join any point T with A, draw the indefinite radius vector AM perpendicular to AT. Make AD $=$ arc PN; join D and N, and draw TM parallel to DN, M will be a point of the curve; for by the construction

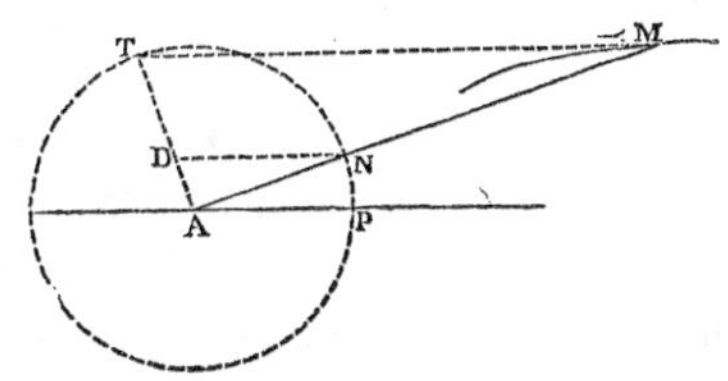

$$AD = \text{tang AND} = \text{tang AMT} = \text{arc NP}.$$

125. The spiral represented by the equation

$$t = \log u$$

is called the *Logarithmic Spiral.*

Differentiating, we find

$$dt = \frac{Mdu}{u};$$

whence

$$\text{tang AMT} = \frac{udt}{du} = M;$$

that is, *the angle formed by the radius vector and tangent is constant,* and the tangent of this angle is equal to the modulus of the system of logarithms used.

If the Naperian system be chosen, $M = 1$, and $AMT = 45°$.

Since t is the logarithm of u, if it be increased uniformly, so that the different arcs t, t', t'', &c., shall be in arithmetical progression, then u, u', u'', &c., must be in geometrical progression, and the curve may be constructed thus. With AO $= 1$ describe a circle, and divide the circumference into any number of equal parts, and draw the lines AO, Ap, Ap', &c. The distances laid off on these lines are to be in geometrical progression, since the arcs Op, Op', Op'', &c., increase by the constant difference Op. To find the ratio of this progression let $t = 0$, then $u = \text{AO} = 1$. Now make $t =$ the arc Op, and find the corresponding value of u in the system of logarithms used, which lay off to m, then

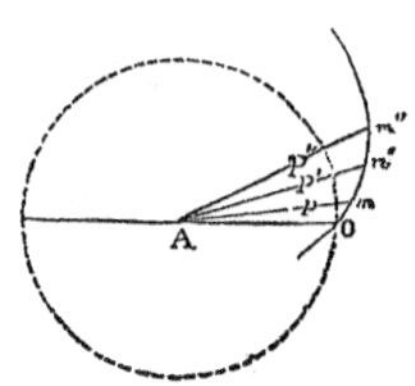

$$\frac{\text{A}m}{\text{AO}} = \text{the ratio.}$$

On Ap', Ap'', &c., lay off Am', Am'', so that

$$\text{AO} : \text{A}m : \text{A}m' : \text{A}m'' : \text{A}m''' : \text{\&c.},$$

m, m', m'', &c., will be points of the curve.

PART II.

INTEGRAL CALCULUS.

---

126. The object of the Integral Calculus is to explain how to pass from differentials to the functions from which they may be derived: Or in any particular case, *to find an expression which, if it be differentiated, will produce the given differential.*

This expression is called *the integral* of the differential. The symbol $\int$ when prefixed to a differential, denotes that its integral is required, thus

$$\int du = u,$$

and this integral (du being infinitely small) is plainly the same as the sum referred to in article (91).

127. We have found, article (15), $dAu = Adu$; therefore

$$\int Adu = \int dAu = Au = A\int du.$$

From which we see that *a constant factor* may be placed without the sign of integration, without affecting the value of the integral; thus,

$$\int b(a - x^2)dx = b\int(a - x^2)dx, \qquad \int\frac{x^3dx}{c} = \frac{1}{c}\int x^3dx.$$

Also in article (18), we have

$$d(u + v \pm \&c.) = du + dv \pm \&c.;$$

hence

$$\int(du + dv \pm \&c.) = \int d(u + v \pm \&c.) = u + v \pm \&c.$$
$$= \int du + \int dv \pm \&c.;$$

that is, *the integral of the sum or difference of any number of differentials, is equal to the sum or difference of their respective integrals.*

Also in article (14), we have

$$d(u + C) = du,$$

no matter what the value of the constant C may be; hence an infinite number of expressions differing from each other in a constant term, when differentiated will produce the same differential. For this reason, *to the integral immediately found we always add a constant*; thus,

$$\int du = u + C.$$

INTEGRATION OF MONOMIAL DIFFERENTIALS, &C.

128. By article (22), we have

$$cdx^{m+1} = c(m + 1)x^m dx;$$

and from this,

$$cx^m dx = \frac{cdx^{m+1}}{m + 1} = cd\frac{x^{m+1}}{m + 1};$$

hence

$$\int cx^m dx = \int cd\frac{x^{m+1}}{m+1} = \frac{cx^{m+1}}{m+1} + C.$$

Therefore, to obtain the integral of a monomial differential: *Multiply the variable with its primitive exponent increased by unity, by the constant factor, if there is one, and divide the result by the new exponent.*

Examples.

1. If $du = xdx$, $\int du = \int xdx = \frac{x^2}{2} + C.$

2. If $du = \frac{x^3 dx}{c}$, $\int du = \frac{1}{c}\int x^3 dx = \frac{x^4}{4c} + C.$

3. If $du = bx^{\frac{2}{3}}dx$, $u = \frac{bx^{\frac{5}{3}}}{\frac{5}{3}} = \frac{3bx^{\frac{5}{3}}}{5} + C.$

4. If $du = \frac{x^{-\frac{m}{n}}dx}{e}$, $u = \frac{nx^{\frac{n-m}{n}}}{e(n-m)} + C.$

5. If $du = \frac{adx}{\sqrt{x}} + \frac{3x^{-\frac{2}{3}}dx}{b} - \frac{c^2x^4dx}{e}$,

$$u = \int \frac{adx}{\sqrt{x}} + \int \frac{3x^{-\frac{2}{3}}dx}{b} - \int \frac{c^2x^4dx}{e} \text{......Art. (127).}$$

The application of the above rule does not give the proper integral when $m = -1$, as in this case we have

$$\int x^{-1}\,dx = \frac{x^{-1+1}}{-1+1} = \frac{1}{0} = \infty,$$

whereas

$$\int x^{-1}dx = \int\frac{dx}{x} = lx + C\text{......Art. (37).}$$

This result was to be expected, since $\int\frac{dx}{x}$ or lx can not be expressed in algebraic terms, Art. (5).

If
$$du = \frac{a}{b}\frac{dx}{x},$$

$$u = \frac{a}{b}\int\frac{dx}{x} = \frac{a}{b}lx + C,$$

or
$$u = \log x + C,$$

the logarithm being taken in the system whose modulus is $\frac{a}{b}$.

129. Many expressions, by the introduction of an auxiliary variable, may be transformed into monomials, and then integrated as in the preceding article.

I. Let

$$du = (a + bx^n)^m c' x^{n-1} dx.$$

Place
$$a + bx^n = z,$$

then

$$nbx^{n-1}dx = dz \qquad x^{n-1}dx = \frac{dz}{bn}.$$

Substituting in the given expression, and integrating, we have

$$\int du = \int \frac{c' z^m dz}{bn} = \frac{c'}{bn}\int z^m dz = \frac{c'}{bn}\,\frac{z^{m+1}}{m+1};$$

and replacing the value of z, we have, finally,

$$u = \frac{c'(a + bx^n)^{m+1}}{(m+1)nb} + C;$$

that is, to integrate a binomial differential when the exponent of the variable without the parenthesis is one less than that within: *Multiply the binomial with its primitive exponent increased by unity, by the constant factor, if there is one, then divide this result by the product of the new exponent, the coefficient and the exponent of the variable within the parenthesis.*

Examples.

1. If $du = (a + bx^2)^{\frac{3}{2}} exdx,$ $\qquad u = \dfrac{e(a + bx^2)^{\frac{5}{2}}}{\frac{5}{2}.b.2} + C.$

2. If $du = (2 - 3x^5)^{-\frac{1}{2}} 3x^4 dx,$ $\quad u = -\dfrac{2}{5}(2 - 3x^5)^{\frac{1}{2}} + C.$

3. If $du = (a - bx^{\frac{2}{3}})^{-\frac{3}{2}} x^{-\frac{1}{3}} dx,$ $\qquad u = \dfrac{3(a - bx^{\frac{2}{3}})^{-\frac{1}{2}}}{b} + C$

4. Let $\qquad du = a(b - cz^{-\frac{m}{n}})^{-\frac{p}{q}} z^{-\frac{m}{n}-1} dz.$

II. Let

$$du = \frac{ax^{n-1}dx}{b \pm x^n}.$$

Place $$b \pm x^n = z;$$

then

$$\pm nx^{n-1}dx = dz \qquad x^{n-1}dx = \pm\frac{dz}{n},$$

and

$$u = \pm\int\frac{adz}{nz} = \pm\frac{a}{n}lz = \pm\frac{a}{n}l(b \pm x^n) + C.$$

In the same way we may find the integrals of the following expressions.

1. Let $$du = \frac{m(b + 2cx)dx}{a + bx + cx^2}.$$

Place $a + bx + cx^2 = z$, then $(b + 2cx)dx = dz$,

$$u = m\int\frac{dz}{z} = mlz = ml(a + bx + cx^2) + C.$$

2. If $du = \dfrac{2dy}{a - y}$, $\qquad u = -2l(a - y) + C.$

3. If $du = \dfrac{(2 + 2x)dx}{2x + x^2}$, $\qquad u = l(2x + x^2) + C.$

4. Let $$du = \frac{2z^{\frac{1}{2}}dz}{1 - z^{\frac{3}{2}}}.$$

Since in general

$$\int\frac{adu}{u} = alu,$$

we see that in all cases where the numerator of an expression is the product of a constant and the differential of the denominator, *its integral will be the product of the constant and the Naperian logarithm of the denominator.*

130. If we have an expression of the form

$$du = (a + bx + cx^2 + \&c.)^m x^n dx,$$

in which m is a positive whole number; the integral may be found by raising the quantity within the parenthesis to the mth power, multiplying each term by $x^n dx$, and then integrating it as in article (128).

Examples.

1. Let $du = (a + x^2)^2 x^3 dx$, or $du = (a^2 + 2ax^2 + x^4)x^3 dx$;

then

$$u = \int (a^2x^3dx + 2ax^5dx + x^7dx) = \frac{a^2x^4}{4} + \frac{2ax^6}{6} + \frac{x^8}{8} + C.$$

2. Let $$du = (b - x^2)^3 x^{\frac{1}{2}} dx.$$

3. Let $$du = (b - cx^{\frac{1}{2}})^2 x^{-\frac{2}{3}} dx.$$

131. Every expression of the form

$$du = Ax^m(a + bx)^n dx,$$

can be integrated, when either m or n is a positive whole number.

If n be positive and entire, we may integrate as in the preceding article.

If m be positive and entire, n being either fractional or negative, place

$$a + bx = z, \qquad \text{then} \qquad x = \frac{z - a}{b},$$

$$dx = \frac{dz}{b}; \qquad du = \mathrm{A}\left(\frac{z - a}{b}\right)^m z^n \frac{dz}{b},$$

$$u = \frac{\mathrm{A}}{b}\int\left(\frac{z - a}{b}\right)^m z^n dz,$$

which may be integrated as in the preceding article. The value of z being then replaced, the integral will be expressed in terms of x.

Examples.

1\. Let $$du = bx^2(a - x)^{\frac{1}{2}}dx.$$

Place $a - x = z$, then $x = a - z$, $dx = -dz$,

$$u = \int - b(a - z)^2 z^{\frac{1}{2}} dz = -\frac{2}{3}ba^2z^{\frac{3}{2}} + \frac{4}{5}baz^{\frac{5}{2}} - \frac{2}{7}bz^{\frac{7}{2}},$$

and finally, by replacing the value of z,

$$u = -\frac{2}{3}ba^2(a - x)^{\frac{3}{2}} + \frac{4}{5}ba(a - x)^{\frac{5}{2}} - \frac{2}{7}b(a - x)^{\frac{7}{2}} + \mathrm{C}.$$

2\. If $$du = \frac{2xdx}{(1 - 3x)^{\frac{1}{2}}},$$

it may be placed under the form

$$du = 2x(1 - 3x)^{-\frac{1}{2}}dx; \quad \text{whence} \quad u = -\frac{2}{9}\int(1 - z)z^{-\frac{1}{2}}dz,$$

and finally,

$$u = -\frac{4}{9}(1 - 3x)^{\frac{1}{2}} + \frac{4}{27}(1 - 3x)^{\frac{3}{2}} + C.$$

3. Let $du = -\dfrac{x^3dx}{1 - x}$. 4. Let $du = \dfrac{ydy}{(3 - 2y)^{\frac{1}{2}}}$.

If

$$du = \frac{(Ax^m + Bx^p + Cx^q + \&c.,)}{(ax + b)^n}dx,$$

we may place it under the form

$$du = \frac{Ax^m dx}{(ax + b)^n} + \frac{Bx^p dx}{(ax + b)^n} + \&c.,$$

and may then integrate each fraction as above, if m, p, q, &c., are entire and positive.

132. To complete each integral as determined by the preceding rules, we have added a constant quantity C. If in the particular case under consideration, we happen to know what the integral must be for a particular value of the variable, this constant can be determined. Thus, if

$$\int X dx = X' + C \ldots\ldots\ldots\ldots(1),$$

X' representing the function of x obtained at once by the application of the rules for integration; and we know the integral must reduce to N when $x = a$, we have

$$N = X'_{x=a} + C, \qquad C = N - X'_{x=a}.$$

In general, however, this constant is entirely arbitrary, since whatever value be assigned to it, it will disappear by differentiation, Art. (14). This arbitrary nature of the constant enables us to cause the integral to fulfil any reasonable condition. Thus if in equation (1), it be required that the integral reduce to the particular expression M, when $x = a$; we may determine the value which must be assigned to C, by writing M for $\int Xdx$, and substituting a for x in the function X′. Calling the result of this substitution A, the equation reduces to

$$M = A + C; \qquad \text{whence} \qquad C = M - A,$$

and

$$\int Xdx = X' + M - A \text{............}(2),$$

which will fulfil the required condition.

If $\quad M = 0, \quad C = -A \quad$ and $\quad \int Xdx = X' - A.$

The integral $\quad \int Xdx = X' + C \quad$ before any particular value has been assigned to C, is called *a complete, or indefinite* integral.

After a particular value has been assigned to C, as in equation (2), it is called *a particular integral;* and if in this particular integral, a particular value be given to x, the result is called *a definite integral.* We should thus have, when $x = b$,

$$\int Xdx = B + M - A \text{............}(3),$$

B representing $\quad X'_{x=b}$.

That value of the variable which causes the integral to reduce to 0 is called *the origin of the integral;* and in every particular integral this origin may be determined by placing the integral equal to 0, and deducing the value of the variable from the resulting equation.

If in (1) we make $\quad x = a, \quad$ and then $\quad x = b, \quad$ we have

$$\int (Xdx)_{x=a} = A + C, \qquad \int (Xdx)_{x=b} = B + C,$$

whence by subtraction,

$$\int (Xdx)_{x=b} - \int (Xdx)_{x=a} = B - A.$$

This is *the integral taken between the limits a and b*, and is usually written

$$\int_a^b Xdx = B - A,$$

the limit corresponding to the subtractive integral being placed below.

If $a, b, c \ldots\ldots\ldots k, l$, be several increasing values of x, and we have

$$\int_a^b Xdx = A', \qquad \int_b^c Xdx = B' \ldots\ldots\ldots \int_k^l Xdx = K';$$

then evidently

$$\int_a^l Xdx = A' + B' + C' \ldots\ldots + K'.$$

Example.

$$\int 6x^2dx = 2x^3 + C;$$

is a complete or indefinite integral.

If it be required that this reduce to 4, when $x = 1$, we have

$$4 = 2 + C, \qquad C = 2,$$

and

$$\int 6x^2dx = 2x^3 + 2, \quad \textit{the particular integral.}$$

For the integral between the limits $x = 0$ and $x = 3$,

$$\int (6x^2dx)_{x=0} = 2, \qquad \int (6x^2dx)_{x=3} = 56;$$

hence

$$\int_0^3 6x^2dx = 54.$$

The origin of the particular integral is obtained by placing

$$2x^3 + 2 = 0; \quad \text{whence} \quad x^3 = -1, \quad x = -1.$$

INTEGRATION OF THE DIFFERENTIALS OF CIRCULAR ARCS.

133. I. In article (42), we have found

$$dx = \frac{du}{\sqrt{1 - u^2}},$$

in which $u = \sin x$, the radius of the circle being unity; then

$$x = \int \frac{du}{\sqrt{1 - u^2}} = \sin^{-1}u + C.$$

Expressions of a similar form may be readily integrated by the aid of an auxiliary variable.

1. Let $$dx = \frac{du}{\sqrt{a^2 - u^2}} \text{......(1)}.$$

Make $u = az$, then $du = adz$, $\sqrt{a^2 - u^2} = a\sqrt{1 - z^2}$. Substituting these values in (1), we have

$$dx = \frac{dz}{\sqrt{1-z^2}}; \qquad x = \sin^{-1} z = \sin^{-1}\frac{u}{a} + C.$$

2. Let
$$dy = \frac{3dx}{\sqrt{2-x^2}}.$$

This may be integrated directly, by placing $x = \sqrt{2}.z$, as in the last example, or by a simple comparison with it, by placing $\sqrt{2}$ for a. Thus

$$y = 3\int \frac{dx}{\sqrt{2-x^2}} = 3 \sin^{-1}\frac{x}{\sqrt{2}} + C.$$

3. Let
$$dy = \frac{2dx}{\sqrt{9-3x^2}}.$$

This should first be placed under the form $dy = \dfrac{2dx}{\sqrt{3}\sqrt{3-x^2}}$.

II. In article (42), we have also

$$dx = -\frac{du}{\sqrt{1-u^2}} = d\cos^{-1}u;$$

whence

$$x = \int -\frac{du}{\sqrt{1-u^2}} = \cos^{-1}u + C.$$

In the same way as in case I, if

$$dx = -\frac{du}{\sqrt{a^2-u^2}}, \qquad x = \cos^{-1}\frac{u}{a} + C.$$

If
$$dx = -\frac{2du}{\sqrt{4-u^2}},$$

$$x = 2\int -\frac{du}{\sqrt{4 - u^2}} = 2\cos^{-1}\frac{u}{2} + C,$$

by placing 4 for a^2.

III. We have also

$$dx = \frac{du}{\sqrt{2u - u^2}} = d \text{ ver-sin}^{-1}u;$$

whence

$$x = \int \frac{du}{\sqrt{2u - u^2}} = \text{ver-sin}^{-1}u + C.$$

1. If
$$dx = \frac{du}{\sqrt{2au - u^2}},$$

place $u = az$, then $du = adz$, and

$$x = \int \frac{du}{\sqrt{2au - u^2}} = \int \frac{dz}{\sqrt{2z - z^2}} = \text{ver-sin}^{-1}z$$

$$= \text{ver-sin}^{-1}\frac{u}{a} + C.$$

2. If

$$dx = \frac{3du}{\sqrt{4u - u^2}}, \qquad x = 3 \text{ ver-sin}^{-1}\frac{u}{2} + C.$$

IV. We have also

$$dx = \frac{du}{1 + u^2} = d \tan^{-1}u;$$

whence

$$x = \int \frac{du}{1 + u^2} = \text{tang}^{-1} u + C.$$

1. If
$$dx = \frac{du}{a^2 + u^2},$$

make $u = az$, then $du = adz$, and

$$x = \int \frac{du}{a^2 + u^2} = \frac{1}{a} \int \frac{dz}{1 + z^2} = \frac{1}{a} \text{tang}^{-1} z = \frac{1}{a} \text{tang}^{-1} \frac{u}{a} + C.$$

2. If $dx = \dfrac{3du}{2 + u^2}$, $\qquad x = \dfrac{3}{\sqrt{2}} \text{tang}^{-1} \dfrac{u}{\sqrt{2}} + C.$

3 Let
$$dy = \frac{2dx}{2 + 3x^2}.$$

If we multiply and divide the fraction $\dfrac{du}{a^2 + u^2}$ by a^2, we have

$$dx = \frac{1}{a^2} \frac{a^2 du}{a^2 + u^2};$$

whence

$$x = \frac{1}{a^2} \int \frac{a^2 du}{a^2 + u^2} = \frac{1}{a^2} \text{tang}^{-1} u + C \ldots\ldots \text{Art. (42)},$$

the radius being a; and in a similar way, all the above expressions may be transformed.

INTEGRATION OF RATIONAL FRACTIONS.

134. Every rational fraction which is the differential of a

function of x, will appear as a particular case of the general form,

$$\frac{(Ax^m + Bx^{m-1} + Cx^{m-2} + \&c.)dx}{A'x^n + B'x^{n-1} + C'x^{n-2} + \&c.}$$

in which m and n are whole numbers and positive.

If m be greater than n, the numerator may be divided by the denominator, and the division continued until the greatest exponent of x in the remainder is one less than in the denominator; the quotient will then consist of an entire and rational part, plus the remainder divided by the denominator, and may be written

$$\frac{Ax^m + Bx^{m-1} + \&c.)dx}{A'x^n + B'x^{n-1} + \&c.} = Xdx + \frac{(A''x^{n-1} + B''x^{n-2} + \&c.)dx}{A'x^n + B'x^{n-1} + \&c.};$$

and the integral of the primitive fraction will be the sum of the integrals of the two parts.

It will be necessary then to explain only the manner of integrating the second part, or those rational fractions in which the greatest exponent of the variable in the numerator is at least one less than in the denominator.

First, suppose the denominator to be divided into its simple factors of the first degree, and let them be represented by

$$x - a, \quad x - b, \quad x - c, \ \&c.$$

There will be four different cases, each of which will require a separate discussion.

1. *When the factors are real and unequal:*
2. *When they are real and equal:*
3. *When they are imaginary and no two alike:*
4. *When they are imaginary and alike, two and two.*

135. I. As an example of the first case, let us take the fraction

$$\frac{(ax + c)dx}{x^2 - b^2}.$$

The two factors of the denominator are $x + b$, and $x - b$; then

$$\frac{(ax + c)dx}{x^2 - b^2} = \frac{(ax + c)dx}{(x + b)(x - b)}.$$

Place $$\frac{ax + c}{x^2 - b^2} = \frac{A}{x + b} + \frac{A'}{x - b} \ldots\ldots\ldots\ldots (1),$$

A and A′ being constants to be determined. For the purpose of determining them, clear the equation of its denominators; then

$$ax + c = Ax - Ab + A'x + A'b.$$

By placing the coefficients of the like powers of x, in the two members, equal to each other, we have

$$a = A + A' \qquad c = A'b - Ab$$

$$A = \frac{ab - c}{2b} \qquad A' = \frac{ab + c}{2b}.$$

Substituting these values in (1), multiplying by dx, and prefixing the sign $\int$, we have

$$\int \frac{(ax + c)dx}{x^2 - b^2} = \frac{ab - c}{2b} \int \frac{dx}{x + b} + \frac{ab + c}{2b} \int \frac{dx}{x - b}$$

$$= \frac{ab - c}{2b} l(x + b) + \frac{ab + c}{2b} l(x - b) + C.$$

The method pursued above indicates the following rule for all similar expressions.

Place the primitive fraction (omitting the differential of the variable), equal to the sum of as many partial fractions as there are factors of the first degree in its denominator; the numerators of these fractions being constants to be determined, and the denominators the several factors of the original denominator; clear the resulting equation of denominators, equate the coefficients of the like powers of the variable in the two members, and thence determine the constants; then multiply each partial fraction by the differential of the variable, and take the sum of their integrals as in case II., *article* (129).

2. Integrate the expression

$$\frac{(3x^2 - 1)}{x^3 - x} dx.$$

The factors of the denominator are, $x + 1$, $x - 1$, and x; then

$$\frac{3x^2 - 1}{x^3 - x} = \frac{A}{x + 1} + \frac{A'}{x - 1} + \frac{A''}{x}.$$

Clearing of denominators,

$$3x^2 - 1 = Ax^2 - Ax + A'x^2 + A'x + A''x^2 - A'';$$

whence

$$3 = A + A' + A'', \qquad 0 = - A + A', \qquad 1 = A'',$$

and

$$A = 1 = A' = A''.$$

Then

$$\int \frac{(3x^2 - 1)}{x^3 - x} dx = \int \frac{dx}{x+1} + \int \frac{dx}{x-1} + \int \frac{dx}{x}$$

$$= l(x+1) + l(x-1) + lx = l(x^3 - x) + C,$$

as may be seen at once, since the numerator of the given differential is the exact differential of the denominator.

3. Integrate the expression

$$\frac{(1-y)dy}{y^2 - 2y - 2}.$$

Placing the denominator equal to 0, we have

$$y^2 - 2y - 2 = 0;$$

whence $y = 1 \pm \sqrt{3}$, and the corresponding factors are

$y - (1 + \sqrt{3})$, $y - (1 - \sqrt{3})$, or $y - m$ and $y - n$.

Finally,

$$\int \frac{(1-y)dy}{y^2 - 2y - 2} = \frac{m-1}{n-m} l(y-m) - \frac{n-1}{n-m} l(y-n) + C.$$

4. Integrate

$$\frac{(2x+3)dx}{x^3 - x^2 - 2x}.$$

5. Integrate

$$\frac{(x^3 - 1)dx}{x^2 - 4}.$$

136 II. In the second case it may be remarked, that if all the factors of the denominator are equal, the fraction will take the form

$$\frac{(\mathrm{A}x^{n-1} + \mathrm{B}x^{n-2} + \&c.,)}{(x - a)^n}dx,$$

which may be integrated as in article (131).

We need then only consider the case where a portion of the factors are equal. The rule of the preceding article is not applicable here, as will be seen by taking the expression

$$\frac{adx}{(x - b)^2(x - c)},$$

in which two of the factors are equal to $x - b$.

By an application of the rule referred to, we should have

$$\frac{a}{(x - b)^2(x - c)} = \frac{\mathrm{A}}{x - b} + \frac{\mathrm{A}'}{x - b} + \frac{\mathrm{A}''}{x - c}$$

$$= \frac{\mathrm{A} + \mathrm{A}'}{x - b} + \frac{\mathrm{A}''}{x - c} = \frac{\mathrm{B}}{x - b} + \frac{\mathrm{A}''}{x - c},$$

since $\mathrm{A} + \mathrm{A}'$ must be regarded as a single constant.

If this equation be cleared of denominators, and the coefficients of the like powers of x in the two members placed equal to each other, we shall evidently form three independent equations, with only two unknown quantities, B and A$''$.

We obviate this difficulty by writing, for the equal factors, the two fractions $\dfrac{\mathrm{B}}{(x - b)^2}$ $\dfrac{\mathrm{B}'}{x - b}$, and thus have

$$\frac{a}{(x - b)^2(x - c)} = \frac{\mathrm{B}}{(x - b)^2} + \frac{\mathrm{B}'}{x - b} + \frac{\mathrm{A}}{x - c},$$

which, being cleared of denominators, gives

$$a = \mathrm{B}(x - c) + \mathrm{B}'(x - b)(x - c) + \mathrm{A}(x - b)^2;$$

whence

$$\mathrm{B}' + \mathrm{A} = 0,\ \mathrm{B} - \mathrm{B}'c - \mathrm{B}'b - 2\mathrm{A}b = 0,\ \mathrm{B}'bc - \mathrm{B}c + \mathrm{A}b^2 = a,$$

three equations with three unknown quantities, which can then be determined.

And in general if there be n equal factors, we should write n partial fractions of the form

$$\frac{\mathrm{B}}{(x - b)^n} + \frac{\mathrm{B}'}{(x - b)^{n-1}} \cdots\cdots \frac{\mathrm{B}^{(n-1)\prime}}{x - b},$$

the numerators of which are constants, and the denominators the different powers of the equal factor from the nth down to the first power. After B, B′, &c. are determined, each partial fraction, being first multiplied by the differential of the variable, will be integrated as in article (129).

Examples.

1. Integrate $\qquad \dfrac{(2 + x)dx}{(x - 1)^2(x - 2)}.$

Place

$$\frac{2 + x}{(x - 1)^2(x - 2)} = \frac{\mathrm{B}}{(x - 1)^2} + \frac{\mathrm{B}'}{x - 1} + \frac{\mathrm{A}}{x - 2}.$$

Clearing of denominators, and equating the coefficients of the like powers of x, we have

$$0 = \mathrm{B}' + \mathrm{A},\quad 1 = \mathrm{B} - 3\mathrm{B}' - 2\mathrm{A},\quad 2 = -2\mathrm{B} + 2\mathrm{B}' + \mathrm{A},$$

$$\mathrm{B} = -3,\qquad \mathrm{B}' = -4,\qquad \mathrm{A} = 4;$$

and finally

$$\int \frac{(2+x)dx}{(x-1)^2(x-2)} = \frac{3}{x-1} - 4l\,(x-1) + 4l\,(x-2) + C.$$

2. Integrate $$\frac{xdx}{x^3 - x^2 - x + 1}.$$

If there are different sets of equal factors, partial fractions must be written for each set; thus,

$$\frac{2}{(x-1)^2(x+1)^2} = \frac{A}{(x-1)^2} + \frac{A'}{x-1} + \frac{B}{(x+1)^2} + \frac{B'}{x+1}.$$

137. III. We know from the general theory of equations, that imaginary roots are found only in pairs, and that for each pair we must have a factor of the second degree, of such a value, that when placed equal to 0, it will give the imaginary roots. Each pair of roots will always appear as a particular case of the general form

$$x = a \pm \sqrt{-b^2} \ldots\ldots\ldots\ldots (1),$$

and the corresponding factor of the second degree will be

$$x^2 - 2ax + a^2 + b^2 = [x-(a+\sqrt{-b^2})]\,[x-(a-\sqrt{-b^2})].$$

By a comparison of the imaginary factors, in any given case, with these general values, we determine the corresponding values of a and b. Thus, if the factor of the second degree be

$$x^2 - 2x + 5,$$

we place it equal to 0, and find the two roots

$$x = 1 \pm \sqrt{-4};$$

whence, by comparison, $a = 1, \quad b^2 = 4, \quad b = 2.$

Now, in the third case, for each pair of imaginary factors, let a partial fraction be written, of the form

$$\frac{Mx + N}{x^2 - 2ax + a^2 + b^2} = \frac{Mx + N}{(x - a)^2 + b^2}.$$

By clearing of denominators, &c., as in the preceding articles, M and N may be determined. We shall have then to integrate the expression

$$\frac{(Mx + N)dx}{(x - a)^2 + b^2}.$$

For this purpose, make $x - a = z$, then $x = z + a$, $dx = dz$.

Substituting these, the original expression becomes

$$\frac{(Mz + Ma + N)}{z^2 + b^2}dz;$$

or by making $Ma + N = P$, and dividing the expression into two parts,

$$\frac{Mzdz}{z^2 + b^2} + \frac{Pdz}{z^2 + b^2}.$$

The first part may be integrated as in case II., Art. (129). Thus,

$$\int \frac{Mzdz}{z^2 + b^2} = \frac{M}{2} l(z^2 + b^2) = Ml\sqrt{z^2 + b^2} = Ml\sqrt{(x-a)^2 + b^2}.$$

The integral of the second part is

$$P\int\frac{dz}{z^2+b^2}=\frac{P}{b}\text{ tang}^{-1}\frac{z}{b}\text{......Art. (133), case IV.,}$$

or by substituting the values of P and z,

$$\int\frac{Pdz}{z^2+b^2}=\frac{N+Ma}{b}\text{tang}^{-1}\left(\frac{x-a}{b}\right);$$

and finally

$$\int\frac{(Mx+N)dx}{(x-a)^2+b^2}=Ml\sqrt{(x-a)^2+b^2}$$

$$+\frac{N+Ma}{b}\text{ tang}^{-1}\frac{(x-a)}{b}+C\text{......(2).}$$

Take the particular example

$$\frac{(x-1)dx}{x^3+x^2+2x}.$$

The factors of the denominator are x and x^2+x+2, the last being the product of the two factors corresponding to the imaginary roots

$$x=-\frac{1}{2}\pm\sqrt{-\frac{7}{4}},$$

which compared with (1), give $a=-\frac{1}{2}$, $b^2=\frac{7}{4}$, $b=\frac{1}{2}\sqrt{7}$.

Place

$$\frac{x-1}{x^3+x^2+2x}=\frac{A}{x}+\frac{Mx+N}{x^2+x+2}.$$

Clearing of denominators &c., we find

$$A = -\frac{1}{2}, \qquad M = \frac{1}{2}, \qquad N = \frac{3}{2}.$$

Substituting these values of M, N, a and b, in formula (2), observing that $\int \frac{Adx}{x} = \int -\frac{1}{2}\frac{dx}{x} = -\frac{1}{2}lx,$ and reducing, we have

$$\int \frac{(x-1)dx}{x^3+x^2+2x} = -\frac{1}{2}\,lx + \frac{1}{2}\,l\,\sqrt{x^2+x+2}$$

$$+ \frac{5}{2\sqrt{7}}\,\text{tang}^{-1}\left(\frac{x+\frac{1}{2}}{\frac{1}{2}\sqrt{7}}\right) + C.$$

138. IV. In the fourth case, where there are several imaginary factors, alike two and two; those of each pair multiplied together will give the same factor of the second degree, and if there be p such pairs, the denominator will contain a factor of the form

$$(x^2 - 2ax + a^2 + b^2)^p.$$

For this, we write p partial fractions; thus

$$\frac{Mx+N}{[(x-a)^2+b^2]^p} + \frac{M'x+N'}{[(x-a)^2+b^2]^{p-1}} \cdots\cdots \frac{M^{(p-1)'}x + N^{(p-1)'}}{(x-a)^2+b^2}.$$

Clearing of denominators &c. the values of M, N, M′, N′, &c. may be determined as before, and since the several partial fractions, after multiplying by dx, are all of the same form, we have only to explain the mode of integrating any one of them except the last, which is to be integrated as in the preceding article. Take the first

$$\frac{(Mx+N)dx}{[(x-a)^2+b^2]^p},$$

and make $x - a = z$; the fraction then becomes

$$\frac{(\mathrm{M}z + \mathrm{M}a + \mathrm{N})dz}{(z^2 + b^2)^p},$$

or placing $\mathrm{M}a + \mathrm{N} = \mathrm{P}$,

$$\frac{\mathrm{M}zdz}{(z^2 + b^2)^p} + \frac{\mathrm{P}dz}{(z^2 + b^2)^p}.$$

The first part is integrated as in case I., Art. (129). Thus

$$\int \frac{\mathrm{M}zdz}{(z^2 + b^2)^p} = \frac{\mathrm{M}(z^2 + b^2)^{-p+1}}{(-p + 1)2} = \frac{\mathrm{M}}{2(1 - p)\,(z^2 + b^2)^{p-1}}.$$

By means of a formula hereafter to be determined, [Formula D, Art. (151)], we shall find

$$\int \frac{\mathrm{P}dz}{(z^2 + b^2)^p} = f(z) + \mathrm{C}'\int \frac{dz}{z^2 + b^2} = f(z) + \frac{\mathrm{C}'}{b}\,\mathrm{tang}^{-1}\frac{z}{b};$$

then

$$\int \frac{(\mathrm{M}z + \mathrm{P})dz}{(z^2 + b^2)^p} = \frac{\mathrm{M}}{2(1 - p)(z^2 + b^2)^{p-1}} + f(z) + \frac{\mathrm{C}'}{b}\,\mathrm{tang}^{-1}\frac{z}{b} + \mathrm{C},$$

after which, substituting the value of z, we shall obtain the complete integral of the primitive expression.

139. By a review of the preceding discussion, it will be seen that all differentials which are rational fractions can be integrated; provided the factors of the denominator can be discovered; and that the integrals will depend upon one or more of the four forms.

$$\int \frac{dx}{x + a}, \quad \int x^m dx, \quad \int \frac{xdx}{(x^2 + a^2)^p}, \quad \int \frac{dx}{x^2 + a^2}.$$

INTEGRATION BY PARTS.

140. In article (19), we have found

$$duv = udv + vdu\,; \quad \text{whence} \quad uv = \int udv + \int vdu,$$

and

$$\int udv = uv - \int vdu \ldots\ldots (1)\,;$$

from which we see, that the integral of udv can be obtained, whenever we are able to integrate vdu. This method of integrating udv is called, *Integration by parts.*

Examples.

1. Integrate the expression $x^3dx\sqrt{a - x^2}$.

This may be divided into the two factors,

$$x^2 \quad \text{and} \quad xdx\sqrt{a - x^2}.$$

Place $\quad x^2 = u \quad$ and $\quad xdx\sqrt{a - x^2} = dv\,;$

then

$$du = 2xdx, \qquad v = \int xdx\sqrt{a - x^2} = -\frac{(a - x^2)^{\frac{3}{2}}}{3}.$$

Substituting these in formula (1), we have

$$\int udv = -\frac{x^2(a - x^2)^{\frac{3}{2}}}{3} + \int \frac{(a - x^2)^{\frac{3}{2}}}{3} 2xdx\,;$$

and finally

$$\int x^3 dx\sqrt{a - x^2} = -\frac{x^2(a - x^2)^{\frac{3}{2}}}{3} - \frac{2}{15}(a - x^2)^{\frac{5}{2}} + C.$$

2. Integrate $\frac{(1 - x^2)^{\frac{1}{2}} dx}{x^2}$.

Place $(1 - x^2)^{\frac{1}{2}} = u$ and $\frac{dx}{x^2} = dv$;

then

$$\int \frac{(1 - x^2)^{\frac{1}{2}}}{x^2} dx = -\frac{\sqrt{1 - x^2}}{x} - \sin^{-1}x + C.$$

3. Integrate $dx\sqrt{1 - x^2}$.

Place $\sqrt{1 - x^2} = u$, and $dx = dv$;

we then have by formula (1),

$$\int dx\sqrt{1 - x^2} = x\sqrt{1 - x^2} + \int \frac{x^2 dx}{\sqrt{1 - x^2}} \text{..........}(2).$$

If we multiply $dx\sqrt{1 - x^2}$ by $\frac{\sqrt{1 - x^2}}{\sqrt{1 - x^2}}$, we may write

$$\int dx\sqrt{1 - x^2} = \int \frac{dx}{\sqrt{1 - x^2}} - \int \frac{x^2 dx}{\sqrt{1 - x^2}} \text{..........}(3).$$

Adding equations (2) and (3), we have

$$2\int dx\sqrt{1-x^2} = x\sqrt{1-x^2} + \int\frac{dx}{\sqrt{1-x^2}},$$

$$\int dx\sqrt{1-x^2} = \frac{x}{2}\sqrt{1-x^2} + \frac{\sin^{-1}x}{2} + C.$$

4. Integrate $\qquad \dfrac{x^2dx}{(a^2-x^2)^{\frac{3}{2}}}.$

INTEGRATION OF CERTAIN IRRATIONAL DIFFERENTIALS.

141. In the preceding articles, rules have been given, *by which every rational differential may be integrated*, except the case referred to in article (139). It may then be taken for granted, that, in general, *every irrational differential which can be made rational in terms of a new variable, can also be integrated.* Let

$$\frac{ax^{\frac{h}{k}}\,dx}{bx^{\frac{m}{n}} + cx^{\frac{p}{q}}}.$$

be a differential, the irrational parts of which are monomials. Make

$$x = z^{knq}; \qquad \text{then} \qquad x^{\frac{h}{k}} = z^{hnq},$$

$$x^{\frac{m}{n}} = z^{mkq}, \qquad x^{\frac{p}{q}} = z^{knp}, \qquad dx = knqz^{knq-1}\,dz.$$

These values substituted in the given expression, evidently make it rational in terms of z and dz. It may then be integrated, after which the value of z in terms of x must be substituted. We may then enunciate the following rule for the integration of expressions of this kind. *For the variable, substitute a new one, with an exponent equal to the least common multiple of the indices of the radicals; then integrate by the known rules, and substitute in the result the value of the new variable in terms of the primitive.*

Examples.

1. Let $$du = \frac{2x^{\frac{1}{2}} - 3x^{\frac{2}{3}}}{5x^{\frac{1}{6}}}dx \ldots\ldots\ldots\ldots(1).$$

The least common multiple of the denominators or indices being 6, we place

$$x = z^6, \qquad \text{then} \qquad dx = 6z^5dz, \qquad z = x^{\frac{1}{6}}.$$

Substituting in (1), we have

$$du = \frac{(2z^3 - 3z^4)}{5z}6z^5dz = \frac{12}{5}z^7dz - \frac{18}{5}z^8dz,$$

and integrating,

$$u = \frac{12}{40}z^8 - \frac{18}{45}z^9 = \frac{3}{10}x^{\frac{4}{3}} - \frac{2}{5}x^{\frac{3}{2}} + C.$$

2. Let $du = \dfrac{3x^{\frac{1}{2}}dx}{2x^{\frac{1}{2}} - x^{\frac{2}{3}}}$. 3. Let $du = \dfrac{axdx}{b - c\sqrt[3]{x}}$.

142. If the irrational parts are all of the form $(a + bx)^{\frac{p}{q}}$ the expression may be made rational in terms of z, by placing

$$a + bx = z^r,$$

r being the least common multiple of the indices of the radicals. We shall thus have

$$x = \frac{z^r - a}{b} \qquad dx = \frac{rz^{r-1}dz}{b},$$

which substituted in the primitive expression, with the value of $a + bx$, will evidently give a rational result. Take the examples;

1. $$du = \frac{dx}{(1 + x)^{\frac{3}{2}} + (1 + x)^{\frac{1}{2}}}.$$

Place $1 + x = z^2$; then $dx = 2zdz$, $z = (1 + x)^{\frac{1}{2}}$.

These values substituted in (1), give

$$du = \frac{2zdz}{z^3 + z} = \frac{2dz}{1 + z^2};$$

whence

$$u = 2\int\frac{dz}{1+z^2} = 2 \text{ tang}^{-1}z = 2 \text{ tang}^{-1} (1 + x)^{\frac{1}{2}} + \text{C}.$$

2. Integrate the expression

$$du = \frac{xdx}{(1 - x)^{\frac{1}{2}} + (1 - x)^{\frac{1}{4}}}.$$

143. Differentials of the form

$$\text{X}dx\left(\frac{a + bx}{a' + b'x}\right)^{\frac{m}{n}},$$

X being a rational function of x, may be made rational by

placing $\dfrac{a + bx}{a' + b'x} = z^n$, deducing the values of x and dx, and substituting them.

For example, let

$$du = xdx\left(\frac{1 + x}{1 - x}\right)^{\frac{2}{3}} \ldots\ldots\ldots\ldots (1).$$

Place $\dfrac{1 + x}{1 - x} = z^3$, then $x = \dfrac{z^3 - 1}{z^3 + 1}$, $dx = \dfrac{6z^2dz}{(z^3 + 1)^2}$.

These values in(1), give

$$du = \frac{(z^3 - 1)6z^4dz}{(z^3 + 1)^3},$$

which is rational.

144. Every radical of the form $\sqrt{a + bx \pm cx^2}$ can be written thus,

$$\sqrt{c}\sqrt{\frac{a}{c} + \frac{b}{c}x \pm x^2} = \sqrt{c}\sqrt{\alpha + \beta x \pm x^2},$$

after making $\dfrac{a}{c} = \alpha$, and $\dfrac{b}{c} = \beta$.

To render rational a differential, the only irrational part of which is a radical of the above form, it will then only be necessary to find rational values for x, dx, and $\sqrt{\alpha + \beta x \pm x^2}$, in terms of a new variable and its differential.

I. Take the case in which the sign of x^2 is $+$, and place

$$\sqrt{\alpha + \beta x + x^2} = z - x \text{............}(1).$$

Squaring both members, we have

$$\alpha + \beta x = z^2 - 2zx;$$

whence

$$x = \frac{z^2 - \alpha}{\beta + 2z} \text{............}(2).$$

By differentiating this value of x, we obtain

$$dx = \frac{2(z^2 + \beta z + \alpha)dz}{(\beta + 2z)^2} \text{............}(3),$$

and by substituting the value of x in the second member of (1),

$$\sqrt{\alpha + \beta x + x^2} = \frac{z^2 + \beta z + \alpha}{\beta + 2z} \text{............}(4).$$

These values of x, dx, and $\sqrt{\alpha + \beta x + x^2}$, substituted in the primitive differential, will evidently give a rational expression in z and dz. After integrating this, the value of z, taken from (1), must be substituted.

Examples.

1. Let $du = \dfrac{dx}{\sqrt{a + bx + cx^2}} = \dfrac{dx}{\sqrt{c}\sqrt{\alpha + \beta x + x^2}}.$

Substituting for dx and $\sqrt{\alpha + \beta x + x^2}$ their values as found above, and reducing, we have

$$du = \frac{dx}{\sqrt{c}\sqrt{\alpha + \beta x + x^2}} = \frac{2dz}{\sqrt{c}(\beta + 2z)};$$

whence

$$u = \frac{1}{\sqrt{c}} l(\beta + 2z) = \frac{1}{\sqrt{c}} l[\beta + 2(\sqrt{\alpha + \beta x + x^2} + x)] + C \ldots\ldots (5).$$

2. Let $du = \dfrac{dx}{\sqrt{h + c^2x^2}} = \dfrac{dx}{c\sqrt{\dfrac{h}{c^2} + x^2}}$.

By comparison with the similar expression in the preceding example, we see that

$$c = \sqrt{c}, \qquad 0 = \beta, \qquad \frac{h}{c^2} = \alpha.$$

Substituting these values in (5), we deduce

$$u = \int \frac{dx}{c\sqrt{\frac{h}{c^2} + x^2}} = \frac{1}{c} l 2(\sqrt{\frac{h}{c^2} + x^2} + x) + C$$

$$= \frac{1}{c} l 2\left(\frac{\sqrt{h + c^2x^2} + cx}{c}\right) = \frac{1}{c} l \frac{2}{c} + \frac{1}{c} l(\sqrt{h + c^2x^2} + cx) + C;$$

and, finally, after uniting the constant $\frac{1}{c} l \frac{2}{c}$, with the arbitrary constant,

$$u = \int \frac{dx}{\sqrt{h + c^2x^2}} = \frac{1}{c} l(\sqrt{h + c^2x^2} + cx) + C.$$

3. Let $$du = \frac{dx\sqrt{2x + x^2}}{x^2}.$$

Comparing this with formulas (2), (3), and (4), we see that

$$0 = \alpha, \qquad 2 = \beta, \qquad x = \frac{z^2}{2 + 2z};$$

$$dx = \frac{2(z^2 + 2z)dz}{(2 + 2z)^2}, \qquad \sqrt{2x + x^2} = \frac{z^2 + 2z}{2 + 2z};$$

whence

$$du = \frac{(z + 2)^2 dz}{z^2(z + 1)}.$$

4. Let $$du = dx\sqrt{m^2 + x^2}.$$

5. Let $$du = \frac{dx}{x\sqrt{x^2 - ax}}.$$

145. II. If the sign of x^2 be minus, it will be necessary to pursue a different method, and deduce other formulas; for if we write

$$\sqrt{\alpha + \beta x - x^2} = z - x,$$

the second powers of x in the squares of the two members will have contrary signs, and not cancel each other, as in the first case, and therefore the deduced value of x in terms of z will not be rational.

Denoting the roots of the equation $x^2 - \beta x - \alpha = 0$, by δ and δ', we have

$$x^2 - \beta x - \alpha = (x - \delta)(x - \delta'),$$

or, changing the signs,

$$\alpha + \beta x - x^2 = (x - \delta)(\delta' - x), \qquad \sqrt{\alpha + \beta x - x^2} = \sqrt{(x - \delta)(\delta' - x)}.$$

Now, if we make

$$\sqrt{(x - \delta)(\delta' - x)} = (x - \delta)z \ldots\ldots\ldots\ldots (1),$$

square both members, and strike out the common factor $x - \delta$, we have

$$\delta' - x = (x - \delta)z^2, \qquad x = \frac{\delta' + \delta z^2}{1 + z^2},$$

$$x - \delta = \frac{\delta' + \delta z^2}{1 + z^2} - \delta = \frac{\delta' - \delta}{1 + z^2} \ldots\ldots (2).$$

Substituting this value in equation (1), we obtain

$$\sqrt{\alpha + \beta x - x^2} = \sqrt{(x - \delta)(\delta' - x)} = \frac{(\delta' - \delta)z}{1 + z^2}.$$

By differentiating equation (2), we find

$$dx = -\frac{2(\delta' - \delta)z dz}{(1 + z^2)^2}.$$

These values of x, $\sqrt{\alpha + \beta x - x^2}$ and dx, substituted in the primitive expression, will make it rational.

Examples.

1. Let $$du = \frac{dx}{\sqrt{\alpha + \beta x - x^2}}.$$

By substituting the values of dx and $\sqrt{\alpha + \beta x - x^2}$, we obtain

$$du = -\frac{2dz}{1+z^2};$$

$$u = -2\int\frac{dz}{1+z^2} = -2\ \text{tang}^{-1}z + \text{C},$$

and since from equation (1),

$$z = \sqrt{\frac{\delta' - x}{x - \delta}},$$

we have finally

$$u = \int\frac{dx}{\sqrt{\alpha + \beta x - x^2}} = -2\ \text{tang}^{-1}\sqrt{\frac{\delta' - x}{x - \delta}} + \text{C}.$$

If in this we make $\beta = 0 \qquad \alpha = 1$, the expression reduces to

$$u = \int\frac{dx}{\sqrt{1 - x^2}} = \text{C} - 2\ \text{tang}^{-1}\sqrt{\frac{1 - x}{1 + x}},$$

since by placing $x^2 - 1 = 0$, we find

$$x = \pm 1 \quad \text{or} \quad \delta = -1 \qquad \delta' = 1.$$

If we now introduce the condition that the integral shall be 0, when $x = 0$, we have

$$0 = \text{C} - 2\ \text{tang}^{-1}\,1 = \text{C} - \frac{\pi}{2}, \qquad \text{C} = \frac{\pi}{2},$$

and

$$\int\frac{dx}{\sqrt{1 - x^2}} = \frac{\pi}{2} - 2\ \text{tang}^{-1}\sqrt{\frac{1 - x}{1 + x}}.$$

The direct integral of the first member is $\sin^{-1}x$, Art. (133); hence

$$\sin^{-1}x = \frac{\pi}{2} - 2\ \mathrm{tang}^{-1}\sqrt{\frac{1-x}{1+x}}.$$

2. Let $$du = \frac{dx\sqrt{2ax - x^2}}{x^2}.$$

Placing $2ax - x^2 = 0$, we deduce $x = 0$, and $x = 2a$; hence $\delta = 0$ and $\delta' = 2a$. Substituting these in the formulas &c., we have

$$\frac{dx\sqrt{2ax - x^2}}{x^2} = -\frac{2z^2dz}{1+z^2},$$

a simple rational fraction.

3. Let $$du = \frac{xdx}{\sqrt{4x - x^2}}.$$

4. Let $$du = \frac{dx\sqrt{2 - x^2}}{x}.$$

INTEGRATION OF BINOMIAL DIFFERENTIALS.

146. 1. If we have a differential of the form

$$x^{m-1}dx(ax^r + bx^n)^{\frac{p}{q}},$$

x^r (r being supposed less than n), may be taken out of the parenthesis, and for the primitive expression we may write

$$x^{m-1}dx\,x^{\frac{rp}{q}}(a + bx^{n-r})^{\frac{p}{q}} = x^{m+\frac{rp}{q}-1}\,dx(a + bx^{n-r})^{\frac{p}{q}},$$

in which but one of the terms, in the parenthesis, contains the variable x, and the exponent of this variable will be positive.

2. If after this, the exponent of x should be fractional, either within or without the parenthesis, or both, we can substitute for x another variable, with an exponent equal to the least common multiple of the denominators of the given exponents, and thus get rid of the fractions, as in the example

$$x^{\frac{1}{3}}dx(a+bx^{\frac{1}{2}})^{\frac{p}{q}};$$

by making $x = z^6$, we obtain

$$x^{\frac{1}{3}}dx(a+bx^{\frac{1}{2}})^{\frac{p}{q}} = 6z^7dz(a+bz^3)^{\frac{p}{q}},$$

in which the exponents of z are whole numbers. *Hence every binomial differential can be placed under the form*

$$x^{m-1}dx(a+bx^n)^{\frac{p}{q}},$$

in which m and n are whole numbers and n positive.

147. 1. The binomial differential being placed under the proposed form; if $\frac{p}{q}$ is entire and positive, it may be integrated as in article (130); if $\frac{p}{q}$ is entire and negative, we have

$$x^{m-1}dx(a+bx^n)^{-\frac{p}{q}} = \frac{x^{m-1}dx}{(a+bx^n)^{\frac{p}{q}}},$$

which is a rational fraction.

2. If $\frac{p}{q}$ is a fraction, either positive or negative, place

$$a + bx^n = z^q;$$

then

$$(a + bx^n)^{\frac{p}{q}} = z^p\ldots\ldots(1), \qquad x^n = \frac{z^q - a}{b},$$

$$x^m = \left(\frac{z^q - a}{b}\right)^{\frac{m}{n}} \qquad x^{m-1}dx = \frac{1}{n}\left(\frac{z^q - a}{b}\right)^{\frac{m}{n}-1}\frac{qz^{q-1}}{b}dz\ldots\ldots(2).$$

The values (1) and (2), substituted in the primitive expression give

$$x^{m-1}dx(a + bx^n)^{\frac{p}{q}} = \frac{q}{nb}z^{p+q-1}\,dz\left(\frac{z^q - a}{b}\right)^{\frac{m}{n}-1}\ldots\ldots(3),$$

which is rational in terms of z and dz, when $\frac{m}{n}$ *is a whole number.*

Example.

Let

$$du = x^3dx(a - bx^2)^{\frac{3}{2}},$$

in which

$$m - 1 = 3, \quad n = 2, \quad \frac{m}{n} = 2, \quad p = 3, \quad q = 2, \quad b = -b.$$

These values in equation (3), give

$$x^3dx(a - bx^2)^{\frac{3}{2}} = z^4dz\,\frac{(z^2 - a)}{b^2},$$

in which

$$z^2 = a - bx^2.$$

3. If $\frac{m}{n}$ is not a whole number, we may write

$$x^{m-1}dx(a + bx^n)^{\frac{p}{q}} = x^{m-1}dx[x^n(\frac{a}{x^n} + b)]^{\frac{p}{q}}$$

$$= x^{m+\frac{np}{q}-1}\,dx(ax^{-n} + b)^{\frac{p}{q}},$$

and in accordance with the preceding principle this will be rational if

$$\frac{m + \frac{np}{q}}{-n} = -\left(\frac{m}{n} + \frac{p}{q}\right) \quad \textit{is a whole number.}$$

To obtain the proper rational expression in terms of z, we need only make in equation (3),

$$m = m + \frac{np}{q}, \quad n = -n, \quad a = b, \quad b = a.$$

Thus

$$x^{m+\frac{np}{q}-1}\,dx(ax^{-n} + b)^{\frac{p}{q}} = -\frac{q}{na}\,z^{p+q-1}\,dz\left(\frac{z^q - b}{a}\right)^{-\frac{m}{n}-\frac{p}{q}-1} \ldots\ldots(4).$$

Example.

Let

$$du = xdx(a + bx^3)^{\frac{1}{3}},$$

$$m-1=1, \quad n=3, \quad p=1, \quad q=3, \quad \frac{m}{n}+\frac{p}{q}=1.$$

These values in equation (4) give

$$xdx(a+bx^3)^{\frac{1}{3}} = -\frac{z^3dz}{a}\left(\frac{z^3-b}{a}\right)^{-2},$$

in which
$$z^3 = ax^{-3}+b.$$

From what precedes, we see that every binomial differential of the proposed form can be integrated: *if the exponent of the parenthesis is a whole number; if the exponent of the variable without the parenthesis plus unity, divided by the exponent of the variable within, is a whole number;* or *if this quotient, plus the exponent of the parenthesis, is a whole number.*

148. Let us now write p for $\frac{p}{q}$, and then divide the expression

$$x^{m-1}dx(a+bx^n)^p = x^{m-n}x^{n-1}dx(a+bx^n)^p,$$

into the two factors

$$x^{m-n}=u \qquad \text{and} \qquad x^{n-1}dx(a+bx^n)^p = dv;$$

whence

$$du=(m-n)x^{m-n-1}dx, \qquad v=\frac{(a+bx^n)^{p+1}}{(p+1)nb} \text{......Art. (129).}$$

Substituting these values in the formula

$$\int udv = uv - \int vdu \text{............Art. (140),}$$

and making $(a + bx^n) = X$, we have

$$\int x^{m-1}dxX^p = \frac{x^{m-n}X^{p+1}}{(p+1)nb} - \frac{(m-n)}{(p+1)nb}\int x^{m-n-1}dxX^{p+1} \ldots\ldots (1).$$

But since

$$X^{p+1} = X^pX = X^p(a + bx^n) = aX^p + bx^nX^p,$$

$$\int x^{m-n-1}dxX^{p+1} = a\int x^{m-n-1}dxX^p + b\int x^{m-1}dxX^p.$$

Substituting this value in (1), and clearing of denominators,

$$(p+1)nb\int x^{m-1}dxX^p = x^{m-n}X^{p+1}$$
$$- (m-n)\left[a\int x^{m-n-1}dxX^p + b\int x^{m-1}dxX^p\right];$$

transposing, &c., we obtain

$$\int x^{m-1}dxX^p = \frac{x^{m-n}X^{p+1} - a(m-n)\int x^{m-n-1}dxX^p}{b(pn+m)} \ldots\ldots \text{(A)}$$

By a single application of this formula we cause

$$\int x^{m-1}dxX^p \qquad \text{to depend upon} \qquad \int x^{m-n-1}dxX^p,$$

in which the exponent of the variable without the parenthesis is diminished by the exponent of the variable within. By an application of the same formula to $\int x^{m-n-1}dxX^p$, it may be made to depend upon $\int x^{m-2n-1}dxX^p$, and finally, by repeated applications, $\int x^{m-1}dxX^p$ will depend upon the expression

$$a(m - rn)\int x^{m-rn-1}dxX^p,$$

in which r represents the number of times m will contain n. If m is an exact multiple of n, then $m - rn = 0$, the term containing the expression to be integrated disappears, and the integration is complete.

If $pn + m = 0$, the second member of the formula becomes infinite, and it fails to answer the purpose; but in this case $p + \frac{m}{n} = 0$, which, substituted in equation (4) of article (147), gives an expression which may at once be integrated.

149. We may also write

$$\int x^{m-1}dxX^p = \int x^{m-1}dxX^{p-1}X = a\int x^{m-1}dxX^{p-1} + b\int x^{m+n-1}dxX^{p-1}.$$

If now in formula **A** we change m into $m + n$, and p into $p - 1$, we have

$$\int x^{m+n-1}dxX^{p-1} = \frac{x^mX^p - am\int x^{m-1}dxX^{p-1}}{b(pn + m)}.$$

Substituting this value in the preceding equation, and reducing, we obtain

$$\int x^{m-1}dxX^p = \frac{x^mX^p + pna\int x^{m-1}dxX^{p-1}}{pn + m} \text{............} \mathbf{B}$$

by which the primitive expression is made to depend upon another, in which the exponent of the parenthesis is one less than before. By repeated applications, this exponent may be reduced to a fraction less than unity, either positive or negative.

150. The use of the preceding formulas may be illustrated by the example

$$\int x^2dx(a + bx^2)^{\frac{3}{2}}.$$

Place $a + bx^2 = X$, $m = 3$, $n = 2$, $p = \frac{3}{2}$, then from formula $\mathfrak{A}_2$

$$\int x^2 dx X^{\frac{3}{2}} = \frac{xX^{\frac{5}{2}} - a\int dxX^{\frac{3}{2}}}{6b}.$$

Applying formula $\mathfrak{B}$ to the expression $\int dxX^{\frac{3}{2}}$ after making $m = 1$, $n = 2$, $p = \frac{3}{2}$, we have

$$\int dxX^{\frac{3}{2}} = \frac{xX^{\frac{3}{2}} + 3a\int dxX^{\frac{1}{2}}}{4},$$

and by another application

$$\int dxX^{\frac{1}{2}} = \frac{xX^{\frac{1}{2}} + a\int \frac{dx}{X^{\frac{1}{2}}}}{2}.$$

Substituting these values, we have finally

$$\int x^2 dxX^{\frac{3}{2}} = \frac{xX^{\frac{5}{2}}}{6b} - \frac{axX^{\frac{3}{2}}}{24b} - \frac{a^2xX^{\frac{1}{2}}}{16b} - \frac{a^3}{16b}\int \frac{dx}{X^{\frac{1}{2}}}.$$

The expression $\frac{dx}{X^{\frac{1}{2}}} = \frac{dx}{\sqrt{a + bx^2}}$ may be integrated as in Art. (144).

151. If in the primitive expressions, m and p are negative, the effect of the application of formulas $\mathfrak{A}$ and $\mathfrak{B}_2$ would evidently be to increase them numerically. Other formulas are then required.

1. From $\mathfrak{A}$ by transposition and reduction, we find

$$\int x^{m-n-1}dx\mathrm{X}^p = \frac{x^{m-n}\mathrm{X}^{p+1} - b(m+np)\int x^{m-1}dx\mathrm{X}^p}{a(m-n)}$$

If in this we change m into $-m+n$, we have

$$\int x^{-m-1}dx\mathrm{X}^p = -\frac{x^{-m}\mathrm{X}^{p+1} - b(n-m+np)\int x^{-m+n-1}dx\mathrm{X}^p}{am} \ldots\ (\mathfrak{C})$$

by the application of which, $-m$ will be numerically diminished by the number of units in n.

2. From $(\mathfrak{B})$, by transposition and reduction, we find

$$\int x^{m-1}dx\mathrm{X}^{p-1} = \frac{-x^m\mathrm{X}^p + (m+np)\int x^{m-1}dx\mathrm{X}^p}{pna}.$$

If in this we change p into $-p+1$, we obtain

$$\int x^{m-1}dx\mathrm{X}^{-p} = \frac{x^m\mathrm{X}^{-p+1} - (m+n-np)\int x^{m-1}dx\mathrm{X}^{-p+1}}{an(p-1)} \ldots\ldots (\mathfrak{D})$$

in which, the exponent of X is numerically one less than in the primitive expression.

If $p-1=0$, the second member becomes infinite, but in this case $p=1$, and the primitive expression reduces to a rational fraction.

152. Let us illustrate the use of these formulas by the example

$$\int x^{-2}dx(2-x^2)^{-\frac{3}{2}}.$$

Making in $(\mathfrak{C})$, $m=1$, $a=2$, $b=-1$, $n=2$, $p=-\frac{3}{2}$, we have

$$\int x^{-2}dx(2-x^2)^{-\frac{3}{2}} = -\frac{x^{-1}\mathrm{X}^{-\frac{1}{2}}}{2} + \int dx\mathrm{X}^{-\frac{3}{2}} \ldots\ldots\ldots\ldots (1).$$

By formula 𝔇, after making $m = 1$, $n = 2$, $a = 2$, $b = -1$, $p = \frac{3}{2}$, we have

$$\int dx X^{-\frac{3}{2}} = \frac{xX^{-\frac{1}{2}}}{2}$$

Making the proper substitutions in (1), we obtain finally

$$\int x^{-2}dx(2 - x^2)^{-\frac{3}{2}} = -\frac{x^{-1}X^{-\frac{1}{2}}}{2} + \frac{xX^{-\frac{1}{2}}}{2} + C,$$

in which $X = 2 - x^2$.

153. By the aid of formula 𝔇 we are now able to integrate the expression

$$\frac{dz}{(z^2+b^2)^p} = dz(z^2 + b^2)^{-p}\text{............Art. (138).}$$

By making $m = 1$, $x = z$, $a = b^2$, $b = 1$, $n = 2$, we cause $\int \frac{dz}{(z^2+b^2)^p}$ to depend upon the integration of another expression in which the exponent is one less, and by repeated applications, we shall find that the integral will depend upon the expression

$$\int \frac{dz}{z^2 + b^2} = \frac{1}{b} \text{tang}^{-1}\frac{z}{b} + C.$$

154. For the expression

$$\int \frac{x^q dx}{\sqrt{2cx - x^2}},$$

we may write

$$\int x^q dx(2cx - x^2)^{-\frac{1}{2}} = \int x^{q-\frac{1}{2}} dx(2c - x)^{-\frac{1}{2}},$$

to which applying formula **A**, after making

$$m = q + \tfrac{1}{2}, \quad a = 2c, \quad b = -1, \quad p = -\tfrac{1}{2}, \quad n = 1,$$

and recollecting that $x^{q-\frac{1}{2}} = x^{q-1}x^{\frac{1}{2}}$, and $x^{q-\frac{3}{2}} = x^{q-1}x^{-\frac{1}{2}}$, we obtain

$$\int \frac{x^q dx}{\sqrt{2cx - x^2}} = -\frac{x^{q-1}\sqrt{2cx - x^2}}{q}$$

$$+ \frac{(2q-1)c}{q}\int \frac{x^{q-1}dx}{\sqrt{2cx - x^2}} \ldots\ldots \mathbf{D}.$$

By repeated applications of this formula, when q is a whole number, we make the primitive expression depend upon

$$\int \frac{dx}{\sqrt{2cx - x^2}} = \text{ver-sin}^{-1}\frac{x}{c} + \text{C} \ldots\ldots \text{Art. (133)}.$$

INTEGRATION BY SERIES.

155. If it be required to integrate the expression Xdx, X being any function of x; it is often convenient and useful to develope X into a series by any of the known methods, generally by the binomial formula; and then, after multiplying by dx, to integrate each term separately. This is called *integrating by series;* since we thus obtain a series equal to the integral of the given expression, from which, when the series is converging, we can for particular values of the variable deduce the approximate value of the integral.

1. Let us take the example

$$du = \frac{dx}{1+x} = dx(1+x)^{-1}.$$

By the binomial formula, we have

$$(1+x)^{-1} = 1 - x + x^2 - x^3 + \&c.$$

Multiplying by dx, and prefixing the sign $\int$,

$$\int \frac{dx}{1+x} = \int (dx - xdx + x^2dx - x^3dx + \&c.);$$

whence

$$\int \frac{dx}{1+x} = x - \frac{x^2}{2} + \frac{x^3}{3} - \frac{x^4}{4} + \&c. + C,$$

or since

$$\int \frac{dx}{1+x} = l\,(1+x),$$

$$l\,(1+x) = x - \frac{x^2}{2} + \frac{x^3}{3} - \frac{x^4}{4} + \&c. + C.$$

But when $x = 0$ the first member becomes $l\,(1) = 0$; hence $C = 0$ and

$$l\,(1+x) = x - \frac{x^2}{2} + \frac{x^3}{3} - \frac{x^4}{4} + \&c. \ldots\ldots \text{Art. (38)}.$$

2. Let $$du = x^{\frac{1}{2}}(1-x^2)^{\frac{1}{2}}dx.$$

By the binomial formula we have

$$(1 - x^2)^{\frac{1}{2}} = 1 - \frac{x^2}{2} - \frac{x^4}{8} - \frac{x^6}{16} - \&c.$$

Multiplying each term by $x^{\frac{1}{2}}dx$, &c.

$$\int x^{\frac{1}{2}}(1 - x^2)^{\frac{1}{2}}dx = \int (x^{\frac{1}{2}}dx - \frac{x^{\frac{5}{2}}dx}{2} - \frac{x^{\frac{9}{2}}dx}{8} - \&c.,$$

whence

$$\int x^{\frac{1}{2}}(1 - x^2)^{\frac{1}{2}}dx = \frac{2}{3}x^{\frac{3}{2}} - \frac{1}{7}x^{\frac{7}{2}} - \frac{x^{\frac{11}{2}}}{44} - \&c \ldots\ldots + C.$$

3. Let $$du = a^x dx.$$

In article (36), we have found

$$a^x = 1 + \frac{kx}{1} + \frac{k^2x^2}{1.2} + \frac{k^3x^3}{1.2.3} + \&c.;$$

hence

$$\int a^x dx = x + \frac{kx^2}{2} + \frac{k^2x^3}{6} + \frac{k^3x^4}{24} + \&c \ldots\ldots + C,$$

in which $k = la$. If $a = e$, then $k = le = 1$, and

$$\int e^x dx = x + \frac{x^2}{2} + \frac{x^3}{6} + \frac{x^4}{24} + \&c \ldots\ldots + C.$$

4. Let $$du = \frac{dx}{\sqrt{x - x^2}} = \frac{dx}{\sqrt{x}\sqrt{1 - x}}$$

Make $\sqrt{x} = u$; then $dx = 2\sqrt{x}du$, and

$$\frac{dx}{\sqrt{x}\sqrt{1 - x}} = \frac{2du}{\sqrt{1 - u^2}},$$

which may be readily integrated, and we shall obtain

$$\int \frac{dx}{\sqrt{x - x^2}} = 2\sin^{-1}\sqrt{x} = 2\sqrt{x}\left(1 + \frac{x}{2.3} + \frac{3x^2}{2.4.5} + \&c......\right) + C.$$

5. Let $$du = dx\sqrt{2ax - x^2};$$

6. Let $$du = \frac{dx\sqrt{1 - e'^2x^2}}{\sqrt{1 - x^2}}.$$

Developing $\sqrt{1 - e'^2x^2} = (1 - e'^2x^2)^{\frac{1}{2}}$, we have

$$\sqrt{1 - e'^2x^2} = 1 - \frac{1}{2}e'^2x^2 - \frac{1}{2}\frac{1}{4}e'^4x^4 - \&c.;$$

hence

$$\int \frac{dx\sqrt{1 - e'^2x^2}}{\sqrt{1 - x^2}} = \int \left(1 - \frac{1}{2}e'^2x^2 - \frac{1}{2}\frac{1}{4}e'^4x^4 - \&c.\right)\left(\frac{dx}{\sqrt{1 - x^2}}\right).$$

After the multiplication, each term of the second member will be of the form $A\int \frac{x^m dx}{\sqrt{1 - x^2}}$, which by formula $\mathfrak{A}_9$ may be made to depend upon

$$\int \frac{dx}{\sqrt{1 - x^2}} = \sin^{-1}x + C.$$

7. Let $$du = \frac{dx}{\sqrt{(2cx - x^2)(b - x)}} = \frac{dx}{\sqrt{2cx - x^2}\sqrt{b - x}}.$$

If we develope $\frac{1}{\sqrt{b - x}} = (b - x)^{-\frac{1}{2}}$, and multiply by $\frac{dx}{\sqrt{2cx - x^2}}$, each term will be of the form $\frac{Ax^q dx}{\sqrt{2cx - x^2}}$ which may be reduced and integrated as in the preceding article.

156. By the application of the formula for integration by parts, Art. (140), to the expression Xdx, we obtain

$$\int Xdx = Xx - \int xdX \ldots\ldots\ldots\ldots(1),$$

and then to xdX, &c.

$$\int xdX = \int \frac{dX}{dx} xdx = \frac{x^2}{2}\frac{dX}{dx} - \int \frac{x^2}{2}\frac{d^2X}{dx} \ldots\ldots\ldots\ldots(2),$$

$$\int \frac{x^2}{2}\frac{d^2X}{dx} = \int \frac{d^2X}{dx^2}\frac{x^2dx}{2} = \frac{x^3}{2.3}\frac{d^2X}{dx^2} - \int \frac{x^3}{2.3}\frac{d^3X}{dx^2},$$

$$\ldots\ldots\ldots\ldots\&c.$$

Substituting in succession the values above deduced, equation (1) will become

$$\int Xdx = Xx - \frac{dX}{dx}\frac{x^2}{1.2} + \frac{d^2X}{dx^2}\frac{x^3}{1.2.3} - \&c.,$$

a series, expressing the integral of Xdx in terms of X, and its differential coefficients; which has received the name of its distinguished discoverer, John Bernouilli.

157. If in the integral

$$\int Xdx = f(x) = u,$$

we make $x = x + h$, we have

$$(\int X dx)_{x=x+h} = f(x+h) = u';$$

and by Taylor's formula,

$$u' - u = \frac{du}{dx}h + \frac{d^2u}{dx^2}\frac{h^2}{1.2} + \text{\&c.}\ldots\ldots\ldots\ldots(1).$$

But since

$$\int X dx = u, \qquad X dx = du, \qquad \frac{du}{dx} = X,$$

$$\frac{d^2u}{dx^2} = \frac{dX}{dx}, \qquad \frac{d^3u}{dx^3} = \frac{d^2X}{dx^2}, \qquad \text{\&c.}$$

These values substituted in (1) give

$$u' - u = Xh + \frac{dX}{dx}\frac{h^2}{1.2} + \frac{d^2X}{dx^2}\frac{h^3}{1.2.3} + \text{\&c.}$$

If in this series we make $x = a$, $h = b - a$, and denote by A, A', A'', &c., what X, $\frac{dX}{dx}$, $\frac{d^2X}{dx^2}$, &c. become under this supposition, it is plain that what u becomes will represent the value of the integral when $x = a$; what u' becomes, its value when $x = a + b - a = b$; then what $u' - u$ becomes, will be the value of the integral between the limits $x = a$, and $x = b$; whence

$$\int_a^b X dx = A(b-a) + \frac{A'}{1.2}(b-a)^2 + \frac{A''}{1.2.3}(b-a)^3 + \text{\&c.},$$

a series from which the approximate value of a definite integral may be obtained. If $b - a$ is so large, that the series does not converge, or does not converge rapidly enough, then let it be divided into n equal parts, so that

$$b - a = n\alpha,$$

and take the value, first between the limits a and $a + \alpha$, then between $a + \alpha$ and $a + 2\alpha$, &c., and suppose the results to be

$$\left.\begin{array}{l} B\alpha + B'\frac{\alpha^2}{1.2} + B''\frac{\alpha^3}{1.2.3} + \&c., \\ C\alpha + C'\frac{\alpha^2}{1.2} + C''\frac{\alpha^3}{1.2.3} + \&c., \\ D\alpha + D'\frac{\alpha^2}{1.2} + D''\frac{\alpha^3}{1.2.3} + \&c., \end{array}\right\} \ldots\ldots(2),$$

$$\&c.;$$

then by article (132) we have

$$\int_a^b Xdx = (B + C + D + \&c.)\alpha + (B' + C' + \&c.)\frac{\alpha^2}{1.2} +$$

$$\&c.\ldots\ldots\ldots\ldots(3),$$

and as α is arbitrary, the separate series (2) [and of course the final series (3)] may be made to converge as rapidly as we please.

INTEGRATION OF DIFFERENTIALS CONTAINING TRANSCENDENTAL QUANTITIES.

158. But few of these differentials admit of exact integrals. We can, however, by the aid of formulas previously deduced, obtain, by series, their approximate integrals.

By the examination of a few expressions, we will endeavour, as far as possible, to indicate to the pupil the general method to be pursued, and then leave to his ingenuity and industry, its application to the different cases with which he may meet.

159. Take first the expression

$$Xa^x dx,$$

in which X is an algebraic function of x. If we divide it into the two factors X and $a^x dx$, and recollect that

$$a^x l a dx = da^x \text{............Art. (36)},$$

whence

$$a^x dx = \frac{da^x}{la}, \qquad \text{and} \qquad \int a^x dx = \frac{a^x}{la};$$

we shall have from the formula for integration by parts

$$\int Xa^x dx = \frac{Xa^x}{la} - \int \frac{a^x}{la} dX \text{............(1)}.$$

If now we take the successive differentials of X, and place

$$dX = X'dx, \qquad dX' = X''dx, \qquad dX'' = X'''dx, \text{ \&c.},$$

we obtain

$$\int \frac{a^x dX}{la} = \frac{X'a^x}{(la)^2} - \int \frac{a^x}{(la)^2} dX',$$

$$\int \frac{a^x dX'}{(la)^2} = \frac{X''a^x}{(la)^3} - \int \frac{a^x}{(la)^3} dX'',$$

&c.

These values in equation (1) give

$$\int Xa^x dx = a^x \left(\frac{X}{la} - \frac{X'}{(la)^2} \cdots\cdots \frac{\pm X^{n'}}{(la)^{n+1}} \right) \mp \int \frac{a^x dX^{n'}}{(la)^{n+1}} \text{......(2)}.$$

If the function X is of such a nature that one of its differential coefficients X′, X″, &c. is constant, the differential of this will be 0, and the corresponding term

$$\mp\int\frac{a^x dX^{n\prime}}{(la)^{n+1}} = 0.$$

The integral will then be exact.

The expression $$x^n a^x dx,$$

admits of an exact integral when n is entire and positive.

If n be fractional or negative, we write for a^x its development, Art. (36), and then integrate as in Art. (155).

160. Take now the expression

$$X(lx)^n dx.$$

If we divide it into the two factors

$$Xdx = dv, \qquad \text{and} \qquad (lx)^n = u;$$

whence

$$\int Xdx = v = X', \qquad du = n(lx)^{n-1}\frac{dx}{x},$$

and then substitute in the formula of Art. (140), we have

$$\int X(lx)^n dx = X'(lx)^n - n\int X'(lx)^{n-1}\frac{dx}{x}\ldots\ldots(1).$$

By this the integral of the primitive expression is made to depend upon the integral of another similar one, in which the exponent of (lx) is one less than at first.

If then n be entire and positive, after repeated applications of the formula, the exponent of (lx) will become 0, and the expression upon which the integral depends, algebraic.

For a particular case, let

$$X = x^m, \qquad \text{then} \qquad \int x^m dx = \frac{x^{m+1}}{m+1} = X',$$

and this in (1) will give

$$\int x^m (lx)^n dx = \frac{x^{m+1}}{m+1}(lx)^n - \frac{n}{m+1}\int x^m (lx)^{n-1} dx \ldots\ldots(2).$$

If in this we substitute for n, in succession

$$n - 1, \qquad n - 2, \qquad n - 3, \ \&c.,$$

we have

$$\int x^m (lx)^{n-1} dx = \frac{x^{m+1}}{m+1}(lx)^{n-1} - \frac{n-1}{m+1}\int x^m (lx)^{n-2} dx,$$

$$\int x^m (lx)^{n-2} dx = \frac{x^{m+1}}{m+1}(lx)^{n-2} - \frac{n-2}{m+1}\int x^m (lx)^{n-3} dx,$$

$$\ldots\ldots\ldots\ldots\&c.$$

These values in (2) will give a general formula, in which, if n be positive and entire, the last term will be

$$\pm \frac{n(n-1)\ldots2.1}{(m+1)^n}\int x^m (lx)^0 dx = \pm \frac{n(n-1)\ldots1\ x^{m+1}}{(m+1)^{n+1}}.$$

We shall therefore have

$$\int x^m(lx)^n dx = \frac{x^{m+1}}{m+1}\left[(lx)^n - \frac{n(lx)^{n-1}}{m+1} + \ldots \pm \frac{n(n-1)\ldots 1}{(m+1)^n}\right] + C \ldots (3).$$

The sign of the last term will be plus when n is even, and minus when n is odd.

If $m = 1$ and $n = 1$, we have

$$\int xlxdx = \frac{x^2}{2}\left(lx - \frac{1}{2}\right) + C.$$

If $m = 0$ and $n = 1$, we have

$$\int lxdx = x(lx - 1) + C.$$

If $m = -1$, the second member of (3) becomes infinite. In this case the differential becomes

$$(lx)^n \frac{dx}{x}.$$

Making $lx = z$, we have $\frac{dx}{x} = dz$, and

$$\int (lx)^n \frac{dx}{x} = \int z^n dz = \frac{z^{n+1}}{n+1} = \frac{(lx)^{n+1}}{n+1} + C,$$

which is true for all values of n, except when $n = -1$. In this case the expression becomes

$$\frac{dx}{xlx}.$$

Making $lx = z$, we have $\frac{dx}{x} = dz$, and

$$\int \frac{dx}{xlx} = \int \frac{dz}{z} = lz = l(lx) + C.$$

161. Take now the expression

$$Xdx \sin^{-1}x.$$

Place $\quad Xdx = dv, \quad$ and $\quad \sin^{-1}x = u, \quad$ then

$$\int Xdx = v = X' \qquad \text{and} \qquad du = \frac{dx}{(1 - x^2)^{\frac{1}{2}}}.$$

Substituting in the formula of Art. (140), we have

$$\int Xdx \sin^{-1}x = X' \sin^{-1}x - \int \frac{X'dx}{(1 - x^2)^{\frac{1}{2}}}.$$

Thus the integral of the primitive expression is made to depend upon the integral of the algebraic expression $\dfrac{X'dx}{(1 - x^2)^{\frac{1}{2}}}$.

Let $$X = x^n,$$

then

$$\int Xdx = \int x^n dx = \frac{x^{n+1}}{n + 1} = X',$$

and we have

$$\int x^n dx \sin^{-1}x = \frac{x^{n+1}}{n + 1} \sin^{-1}x - \frac{1}{n + 1} \int \frac{x^{n+1}\, dx}{(1 - x^2)^{\frac{1}{2}}}.$$

By the application of formula A or C, when n is entire, the last term may be reduced, and then integrated; except when $n = -1$, in which case the expression becomes

$$\frac{dx}{x} \sin^{-1}x,$$

which can only be integrated by series.

In the same way, like expressions may be found for

$$\int \mathrm{X}dx \cos^{-1}x, \qquad \int \mathrm{X}dx \text{ tang}^{-1}x, \text{ \&c.}$$

162. By article (41) we have

$$d \sin nx = ndx \cos nx, \qquad d \cos nx = -ndx \sin nx;$$

hence

$$\int dx \sin nx = -\frac{\cos nx}{n}, \qquad \int dx \cos nx = \frac{\sin nx}{n}.$$

In the expression

$$dx \sin^2 x,$$

we can place for $\sin^2 x$, its value, $\frac{1}{2} - \frac{\cos 2x}{2}$, and then have

$$\int dx \sin^2 x = \int \frac{dx}{2} - \int \frac{\cos 2xdx}{2} = \frac{x}{2} - \frac{1}{4}\sin 2x + \mathrm{C};$$

and in general the integral of similar expressions containing any power of either the sine or cosine of x, can be obtained by first substituting the value of the power in terms of the double, triple, &c. arc, as determined in trigonometry.

The expressions

$$dx \sin^m x, \qquad dx \cos^m x,$$

when m is entire, may also be integrated as follows. Make

$$\sin x = z, \qquad \text{then} \qquad x = \sin^{-1} z \qquad dx = \frac{dz}{(1-z^2)^{\frac{1}{2}}};$$

whence

$$\int dx \sin^m x = \int \frac{z^m dz}{(1 - z^2)^{\frac{1}{2}}}.$$

This expression, by repeated applications of formula **A** or **C**, may be made to depend upon

$$\int \frac{dz}{(1 - z^2)^{\frac{1}{2}}}, \qquad \text{or} \qquad \int \frac{z dz}{(1 - z^2)^{\frac{1}{2}}}.$$

In the expression

$$dx \tan^m x,$$

place

$$\tan x = z$$

then

$$dx = \frac{dz}{1 + z^2}, \qquad \int dx \tan^m x = \int \frac{z^m dz}{1 + z^2},$$

which is a rational fraction.

Examples.

Integrate 1. $du = dx \sin^3 x$. 2. $du = \dfrac{dx}{\cos^3 x}$.

3. $du = \dfrac{dx}{\sin x}$. 4. $du = dx \tan^2 x$.

163. In the general expression

$$dx \sin^m x \cos^n x,$$

we may place

$$\sin x = z, \quad \text{then} \quad \cos x = (1 - z^2)^{\frac{1}{2}}, \quad dx = \frac{dz}{(1-z^2)^{\frac{1}{2}}},$$

and finally,

$$\int dx \sin^m x \cos^n x = \int z^m dz (1 - z^2)^{\frac{n-1}{2}},$$

which may be reduced by formulas **A**, **B**, **C**, and **D**, and in some cases integrated, as in the example

$$du = dx \sin^4 x \cos^2 x; \quad \text{whence} \quad u = \int z^4 dz (1 - z^2)^{\frac{1}{2}}.$$

INTEGRATION OF DIFFERENTIALS OF THE HIGHER ORDERS.

164. By an application of the rules previously demonstrated, we may readily obtain the primitive function, from which differentials, containing a single variable, and of a higher order than the first, may have been derived.

Let there be the differential

$$d^n u = f(x) dx^n.$$

Dividing by dx^{n-1}, we have

$$\frac{d^n u}{dx^{n-1}} = f(x) dx,$$

and since dx^{n-1} is a constant, this may be written, Art. (24),

$$d\left(\frac{d^{n-1}u}{dx^{n-1}}\right) = f(x)dx.$$

Integrating both members, we have

$$\frac{d^{n-1}u}{dx^{n-1}} = \int f(x)dx = f'(x) + \text{C}.$$

After multiplying both members of this equation by dx, it may be written

$$d\left(\frac{d^{n-2}u}{dx^{n-2}}\right) = f'(x)dx + \text{C}dx;$$

and integrating as before,

$$\frac{d^{n-2}u}{dx^{n-2}} = f''(x) + \text{C}x + \text{C}';$$

which by another transformation and integration, may be reduced one degree lower, and finally after n integrations, we shall obtain

$$u = \text{F}(x) + \frac{\text{C}x^{n-1}}{1.2\ldots(n-1)} + \frac{\text{C}'x^{n-2}}{1.2\ldots(n-2)} + \ldots\ldots\ldots\ldots \text{C}^{(n-1)\prime}.$$

The above operation may be indicated thus,

$$u = \int{}^{n} f(x)dx^{n};$$

the symbol $\int{}^{n}$ indicating that n successive integrations are required.

Examples.

1. Let $$d^2u = ax^2dx^2.$$

The required operation is indicated thus,

$$u = \int^2 ax^2dx^2,$$

and may be read, *the double integral of* ax^2dx^2.

Let the expression, after dividing by dx, be written

$$\frac{d^2u}{dx} = d\left(\frac{du}{dx}\right) = ax^2dx;$$

whence by integration

$$\frac{du}{dx} = \frac{ax^3}{3} + \text{C}, \qquad du = \frac{ax^3}{3}dx + \text{C}dx.$$

Integrating again, we obtain

$$u = \frac{ax^4}{12} + \text{C}x + \text{C}'.$$

2. If

$$d^3u = bdx^3, \qquad u = \int^3 bdx^3,$$

which is called *a triple integral.* We may write

$$\frac{d^3u}{dx^2} = d\left(\frac{d^2u}{dx^2}\right) = bdx;$$

whence

$$\frac{d^2u}{dx^2} = bx + C,$$

and finally as in the last example

$$u = \int^3 bdx^3 = \frac{bx^3}{6} + \frac{Cx^2}{2} + C'x + C''.$$

3 Let $\qquad d^4u = -\frac{6dx^4}{x^4}. \qquad$ 4. Let $\qquad d^3u = \sqrt{x}dx^3.$

INTEGRATION OF PARTIAL DIFFERENTIALS.

165. Hitherto, we have explained the mode of integrating only the differentials of functions of a single variable. It yet remains to extend our rules to the integration of those which contain more than one variable.

These differentials are either *partial* or *total*, Art. (49). When partial, they belong to one of *two classes*.

I. Those obtained from the primitive function by differentiating with reference to one variable only.

II. Those obtained by differentiating first with reference to one variable, and then with reference to another, &c., Art. (46).

166. The differentials of the first class may be expressed generally thus,

$$d^nu = f(x, y, z, \&c.)\,dx^n \qquad d^nu = f'(x, y, z, \&c.)dy^n, \ \&c.,$$

in which u is a function of x, y, z, &c., and may evidently be obtained by successive integrations, precisely as in article (164); all the variables, except the one with reference to which the differentiation was made, being regarded as constant, and care being taken to add, instead of constants, arbitrary functions of those variables which are regarded as constant during the integration.

Examples.

1. Let $$d^2u = bx^2ydx^2,$$

which, after dividing by dx, may be written

$$d\left(\frac{du}{dx}\right) = bx^2ydx;$$

whence

$$\frac{du}{dx} = \int bx^2ydx = \frac{bx^3y}{3} + \text{Y},$$

$$du = \frac{bx^3y}{3}dx + \text{Y}dx,$$

and

$$u = \int^2 bx^2ydx^2 = \frac{bx^4y}{12} + \text{Y}x + \text{Y}',$$

in which Y and Y′ are arbitrary functions of y.

2. Let $$d^3u = cx^2y^2z^2dy^3.$$

167. The differentials of the second class may be written generally thus,

$$d^{m+n+\cdots}u = f(x, y, z, \&c.)dx^m\,dy^n\,dz^p\ldots\ldots,$$

and the mode of integrating is plainly to integrate first, m times with reference to x, then n times with reference to y, and so on until all the required integrations are made.

To illustrate, let

$$d^2u = \varphi(x, y)dxdy,$$

which may be written

$$\frac{d^2u}{dx} = \varphi(x, y)dy, \qquad \text{or} \qquad d\left(\frac{du}{dx}\right) = \varphi(x, y)dy;$$

whence by integration with reference to y,

$$\frac{du}{dx} = \int \varphi(x, y)dy + X, \qquad du = dx\int \varphi(x, y)dy + Xdx,$$

and

$$u = \int dx \int \varphi(x, y)dy + \int Xdx + Y,$$

or

$$u = \int^2 \varphi(x, y)dydx + \int Xdx + Y,$$

there being no necessity of indicating with reference to which variable the integration is first to be made, Art. (47).

Examples.

1. Let $$d^3u = ax^2ydy^2dx.$$

This may be written

$$\frac{d^3u}{dy^2} = ax^2ydx, \qquad \text{or} \qquad d\left(\frac{d^2u}{dy^2}\right) = ax^2ydx.$$

Integrating with reference to x,

$$\frac{d^2u}{dy^2} = \frac{ax^3y}{3} + Y,$$

which may now be integrated as in the preceding article.

2. Let $\qquad d^3u = axz^2dxdydz.$

3. Let $\qquad d^4u = (x + y)^2dx^2dy^2.$

INTEGRATION OF TOTAL DIFFERENTIALS OF THE FIRST ORDER.

168. If $\qquad u = f(x, y),$

we have found, Art. (49),

$$du = \frac{du}{dx}dx + \frac{du}{dy}dy,$$

in which, $\frac{du}{dx}dx$ and $\frac{du}{dy}dy$ are the partial differentials of $f(x, y)$; and also, Art. (47),

$$\frac{d^2u}{dxdy} = \frac{d^2u}{dydx} \quad \text{or} \quad \frac{d\left(\frac{du}{dx}\right)}{dy} = \frac{d\left(\frac{du}{dy}\right)}{dx} \ldots\ldots(1).$$

If then an expression of the form

$$Pdx + Qdy \ldots\ldots(2)$$

be the total differential of a function of x and y; Pdx and Qdy

must be the two partial differentials of the function, and by the integration of either, we shall obtain the function itself.

To ascertain, in any given expression of the above form, whether Pdx and Qdy are such partial differentials, we have simply to see if the condition (1), or

$$\frac{dP}{dy} = \frac{dQ}{dx},$$

is fulfilled. If so, the given expression is the differential of a function of x and y, and we have

$$u = \int Pdx + Y \ldots\ldots (3),$$

Y being a function of y, which is to be determined so as to satisfy the condition $\frac{du}{dy} = Q$.

To determine this value of Y, let equation (3) be differentiated with reference to y. Then

$$\frac{du}{dy} = \frac{d\int Pdx}{dy} + \frac{dY}{dy};$$

or representing $\int Pdx$ by v,

$$\frac{du}{dy} = \frac{dv}{dy} + \frac{dY}{dy} = Q;$$

whence

$$\frac{dY}{dy} = Q - \frac{dv}{dy}, \quad Y = \int\left(Q - \frac{dv}{dy}\right)dy,$$

and finally

$$u = \int Pdx + \int\left(Q - \frac{dv}{dy}\right)dy.$$

Examples.

1. Let

$$du = (2axy - 3bx^2y)dx + (ax^2 - bx^3)dy,$$

which compared with equation (2), gives

$$P = 2axy - 3bx^2y, \qquad Q = ax^2 - bx^3,$$

$$\frac{dP}{dy} = 2ax - 3bx^2 = \frac{dQ}{dx}.$$

This condition being fulfilled, we then have

$$u = \int (2axy - 3bx^2y)dx = ax^2y - byx^3 + Y.$$

To determine Y, we have

$$v = \int Pdx = ax^2y - byx^3,$$

and

$$\frac{dv}{dy} = ax^2 - bx^3; \qquad \text{whence} \qquad Q - \frac{dv}{dy} = 0, \ Y = C.$$

2. If $\qquad du = \dfrac{dx}{y} + \left(2y - \dfrac{x}{y^2}\right)dy,$

$$u = \int \frac{dx}{y} = \frac{x}{y} + Y.$$

Since $\quad v = \int Pdx = \dfrac{x}{y}, \quad$ we have

$$\frac{dv}{dy} = -\frac{x}{y^2};$$

hence

$$Y = \int\left(Q - \frac{dv}{dy}\right)dy = \int 2ydy = y^2 + C,$$

and

$$u = \frac{x}{y} + y^2 + C.$$

3. If $\qquad du = \dfrac{ydx - xdy}{x^2 + y^2}, \qquad u = \tan^{-1}\dfrac{x}{y} + C.$

4. Let $\qquad du = (6xy - y^2)dx + (3x^2 - 2xy)dy.$

169. If a function of two variables, composed of entire terms, is homogeneous with reference to them, its differential will also be homogeneous; and such a relation will exist between the function and its partial differential coefficients, as will enable us at once to obtain the function, when the differential is given.

To explain this relation, let

$$u = f(x, y),$$

and m denote the sum of the exponents of x and y in each term. For x and y substitute tx and ty respectively, the primitive function then becomes $t^m u$.

In this expression, for t put $(1 + s)$; then

$$t^m u = (1 + s)^m u.$$

Under these suppositions, x and y, in the primitive function, have become, respectively, $x + sx$, and $y + sy$.

Developing this new state of the primitive function, as in article (46), we have

$$u + \left(\frac{du}{dx}sx + \frac{du}{dy}sy\right) + \frac{1}{2}\left(\frac{d^2u}{dx^2}s^2x^2 + 2\frac{d^2u}{dxdy}s^2xy\ldots\right) + \&c.$$

$$= (1+s)^m u = u + mus + \frac{m(m-1)us^2}{1.2} + \&c.\ldots\ldots$$

Equating the coefficients of the first powers of the indeterminate s, we have

$$\frac{du}{dx}x + \frac{du}{dy}y = mu\ldots\ldots(1).$$

Hence in the differential

$$du = \mathrm{P}dx + \mathrm{Q}dy,$$

if P and Q are homogeneous of the $(m-1)$th degree, we shall have, by comparison with equation (1),

$$\mathrm{P}x + \mathrm{Q}y = mu; \qquad u = \frac{\mathrm{P}x + \mathrm{Q}y}{m}.$$

For example, let

$$du = 4xy^2dx + ay^3dx + 4x^2ydy + 3axy^2dy,$$

in which, $m - 1 = 3$, $m = 4$,

$$4xy^2 + ay^3 = \mathrm{P}, \qquad 4x^2y + 3axy^2 = \mathrm{Q};$$

whence

$$u = \frac{\mathrm{P}x + \mathrm{Q}y}{4} = 2x^2y^2 + axy^3.$$

170. The method of obtaining the integral of a differential, containing several variables, is readily deduced from what precedes. Let

$$du = \mathrm{P}dx + \mathrm{Q}dy + \mathrm{R}dz = df(x, y, z) \ldots\ldots\ldots\ldots (1).$$

If for a moment we regard z as a constant, and then in succession y and x, it is plain that we shall have the three expressions

$$\mathrm{P}dx + \mathrm{Q}dy, \qquad \mathrm{P}dx + \mathrm{R}dz, \qquad \mathrm{Q}dy + \mathrm{R}dz \ldots\ldots (2),$$

which, taken separately, are differentials of functions of two variables, if the primitive expression is a differential of a function of three, and the reverse.

But the conditions that these be each an exact differential, are

$$\frac{d\mathrm{P}}{dy} = \frac{d\mathrm{Q}}{dx}, \qquad \frac{d\mathrm{P}}{dz} = \frac{d\mathrm{R}}{dx}, \qquad \frac{d\mathrm{Q}}{dz} = \frac{d\mathrm{R}}{dy} \ldots\ldots (3);$$

hence if we have given an expression of the form

$$\mathrm{P}dx + \mathrm{Q}dy + \mathrm{R}dz,$$

and the conditions (3) are fulfilled, it will be the differential of a function of three variables, and we can obtain the function by integrating either of the expressions (2), as in Art. (168), taking care to add to the integral a function of that variable which is regarded as constant. Thus, denoting the integral of
$\mathrm{P}dx + \mathrm{Q}dy$ by v, we have

$$\int (\mathrm{P}dx + \mathrm{Q}dy + \mathrm{R}dz) = v + \mathrm{Z} \ldots\ldots\ldots (4),$$

Z being independent of x and y, and a function of z alone.

If now we differentiate equation (4) with reference to z, we find

$$R = \frac{dv}{dz} + \frac{dZ}{dz};$$

whence

$$\frac{dZ}{dz} = R - \frac{dv}{dz}; \qquad Z = \int\left(R - \frac{dv}{dz}\right)dz + C,$$

and finally

$$u = \int(Pdx + Qdy + Rdz) = v + \int\left(R - \frac{dv}{dz}\right)dz + C.$$

By a similar course of reasoning, we may deduce the integral of the differential of a function of any number of variables.

171. In article (168) we have denoted $\int Pdx$ by v; whence

$$\frac{dv}{dx} = P.$$

Differentiating this with reference to the variable y, we find

$$\frac{d\left(\frac{dv}{dx}\right)}{dy} = \frac{dP}{dy} = \frac{d\left(\frac{dv}{dy}\right)}{dx} \text{......Art. (168);}$$

whence

$$\frac{d\left(\frac{dv}{dy}\right)}{dx}dx = \frac{dP}{dy}dx.$$

Integrating with reference to the variable x, we have

$$\frac{dv}{dy} = \int \frac{dP}{dy} dx,$$

or since $(dP)dx = d(Pdx)$,

$$\frac{d \int Pdx}{dy} = \int \frac{d(Pdx)}{dy}.$$

By which we see *that we may differentiate with reference tc another variable, the indicated integral of a partial differential, by simply differentiating the quantity under the sign.*

INTEGRATION OF DIFFERENTIAL EQUATIONS.

172. These equations when of the first order, and when derived from equations containing but two variables, will appear as particular cases of the general form

$$Pdx + Qdy = 0,$$

and may of course be integrated as in article (168), when

$$\frac{dP}{dy} = \frac{dQ}{dx},$$

and give

$$\int Pdx + Y = C.$$

In practice, however, it will in general be found, that in consequence of the disappearance of a factor, common to both terms of the differential equation, or when the differential equation has been obtained by the elimination of a constant from the primitive and its immediate differential equation, Art. (56), this condition is not fulfilled; hence other means of obtaining the integral must be sought for.

In the first place, it is evident that, if by any transformation the equation can be placed under the form

$$Xdx + Ydy = 0,$$

X being a function of x and Y of y, the integral can be found by taking the sum of the integrals of the two terms; thus

$$\int Xdx + \int Ydy = C.$$

173. Among the most simple forms with which we meet, are

I. $$Ydx + Xdy = 0.$$

II. $$XYdx + X'Y'dy = 0.$$

The variables may be separated, in I. by dividing by YX, and in II. by dividing by YX'. The results

$$\frac{dx}{X} + \frac{dy}{Y} = 0,$$

and

$$\frac{X}{X'}dx + \frac{Y'}{Y}dy = 0,$$

are under the proposed form. In general, if the value of $\frac{dy}{dx}$, deduced from the equation, be under the form

$$\frac{dy}{dx} = XY,$$

we have

$$\frac{dy}{Y} = Xdx; \quad \text{and} \quad \int \frac{dy}{Y} = \int Xdx.$$

Examples.

1. Let $$ydx - xdy = 0.$$

Dividing by yx, we have

$$\frac{dx}{x} - \frac{dy}{y} = 0, \qquad lx - ly = C,$$

or making $C = lC'$, we have

$$l\frac{x}{y} = lC', \qquad \frac{x}{y} = C' \qquad x = C'y.$$

2. Let $$xy^2dx + dy = 0.$$

Dividing by y^2,

$$xdx + \frac{dy}{y^2} = 0;$$

integrating, and reducing

$$x^2y - 2 = 2Cy.$$

3. Let $$(1 - x)^2ydx - (1 + y)x^2dy = 0;$$

whence

$$\frac{(1-x)^2}{x^2}dx - \frac{1+y}{y}dy = 0,$$

and

$$-\frac{1}{x} - 2lx + x - ly - y = C.$$

4. Let $$(1 + x^2)dy - \sqrt{y}\,dx = 0.$$

5. Let $$x^2ydx - (3y + 1)\sqrt{x^3}dy = 0.$$

174. III. In all cases where the equation is homogeneous with reference to the variables, they can be separated, and the equation placed under the proposed form.

Suppose the general form of the given differential to be

$$Ax^ny^mdx + Bx^hy^kdy = 0,$$

in which $$n + m = h + k = g.$$

Make $y = zx$, and substitute; we thus obtain

$$Ax^gz^mdx + Bx^gz^kdy = 0;$$

dividing by x^g, and putting for dy its value, $zdx + xdz$; we have

$$Az^mdx + Bz^k(zdx + xdz) = 0;$$

dividing by $(Az^m + Bz^{k+1})x$, we have

$$\frac{dx}{x} + \frac{Bz^kdz}{Az^m + Bz^{k+1}} = 0,$$

which is under the proposed form.

Examples.

1. Let $$x^2dy - y^2dx - xydx = 0.$$

Make $y = zx$, then $dy = zdx + xdz$.

Substituting in the given equation, we have

$$x^2zdx + x^3dz - z^2x^2dx - x^2zdx = 0;$$

reducing and integrating,

$$xdz - z^2dx = 0, \qquad -\frac{1}{z} - lx = C.$$

Putting for z its value, we have finally

$$lx = -(C + \frac{x}{y}).$$

2. If $\frac{x^2 + yx}{x - y}dy = ydx, \qquad lx = \frac{x}{2y} - l\sqrt{\frac{y}{x}} + C.$

3. Let $\qquad xdy - ydx = dx\sqrt{x^2 + y^2}.$

175. IV. The equation

$$(a + bx + cy)dx + (a' + b'x + c'y)dy = 0,$$

may be so transformed, that the variables can be separated and the integral found. For this purpose let us make

$$x = t + \delta \qquad y = u + \delta';$$

whence

$$dx = dt, \qquad dy = du.$$

These values in the primitive equation, give

$$(a + b\delta + c\delta' + bt + cu)dt + (a' + b'\delta + c'\delta' + b't + c'u)du = 0.$$

By placing

$$a + b\delta + c\delta' = 0, \qquad a' + b'\delta + c'\delta' = 0,$$

we can determine proper values for the arbitrary quantities δ and δ', and our equation reduces to

$$(bt + cu)dt + (b't + c'u)du = 0;$$

which being homogeneous with reference to t and u may be treated as in the preceding article.

This transformation is always possible, save when the values of δ and δ' become infinite, which will be the case only when

$$bc' - cb' = 0;$$

whence

$$c' = \frac{cb'}{b}; \qquad b'x + c'y = \frac{b'}{b}(bx + cy).$$

The primitive equation thus becomes

$$adx + a'dy + (bx + cy)(dx + \frac{b'}{b}dy) = 0,$$

in which the variables may be separated by making

$$bx + cy = z.$$

Substituting this, and the resulting value of dy, the equation reduces to

$$dx + \frac{(a'b + b'z)dz}{abc - a'b^2 + (bc - bb')z} = 0.$$

If $$bc - bb' = 0,$$

we have at once the integral

$$x + \frac{2a'bz + b'z^2}{2(abc - a'b^2)} = C,$$

in which the value of z is to be substituted.

176. V. In the equation

$$dy + Pydx = Qdx \ldots\ldots\ldots\ldots (1),*$$

P and Q being functions of x, the variables may be readily separated by making

$$y = zX \ldots\ldots\ldots\ldots (2),$$

X being a function of x, for which a proper value is to be determined. By differentiating equation (2), we have

$$dy = zdX + Xdz,$$

and by substitution in (1),

$$zdX + X(dz + Pzdx) = Qdx \ldots\ldots\ldots\ldots (3).$$

Suppose X to have such a value that

$$zdX = Qdx \ldots\ldots\ldots\ldots (4);$$

equation (3) then becomes

$$X(dz + Pzdx) = 0;$$

whence

$$\frac{dz}{z} = -Pdx; \qquad lz = -\int Pdx,$$

or taking the numbers

$$z = e^{-\int Pdx}.$$

---

* NOTE.—Equations of this kind being of the first degree with reference to y and dy, are sometimes improperly called *linear equations.*

From equation (4), we have

$$dX = \frac{Qdx}{z} = Qe^{\int Pdx}dx;$$

whence

$$X = \int Qe^{\int Pdx}dx.$$

These values of z and X, in equation (2), give

$$y = e^{-\int Pdx}\int Qe^{\int Pdx}dx.$$

177. Equations of the form

$$ay^m x^n dx + bx^p dx + cx^q dy = 0,$$

may sometimes be rendered homogeneous by making

$$y = z^k,$$

k being a constant to be determined. From this, we have

$$dy = kz^{k-1}dz, \qquad y^m = z^{km}.$$

These values in the primitive equation give

$$az^{km}x^n dx + bx^p dx + ckx^q z^{k-1}dz = 0,$$

which will be homogeneous, if

$$km + n = p = q + k - 1,$$

that is, when

$$\frac{p-n}{m} = p + 1 - q = k.$$

178. It has been remarked, article (172), that differential equations sometimes fail to fulfil the condition of integrability, in consequence of the disappearance of a common factor. Whenever this factor can be discovered, by trial or otherwise, the integral can at once be found, as in article (168).

Let

$$Pdx + Qdy = 0$$

be a differential equation, in which the condition is not fulfilled, and suppose that

$$z = f(x, y)$$

is the factor by the disappearance of which the given equation has resulted. The immediate differential equation will then be

$$Pzdx + Qzdy = 0,$$

from which we have the condition

$$\frac{dPz}{dy} = \frac{dQz}{dx},$$

or performing the differentiation

$$\frac{zdP}{dy} + \frac{Pdz}{dy} = \frac{zdQ}{dx} + \frac{Qdz}{dx},$$

or

$$\left(P\frac{dz}{dy} - Q\frac{dz}{dx}\right) + \left(\frac{dP}{dy} - \frac{dQ}{dx}\right)z = 0 \ldots\ldots(1).$$

This equation expresses a relation between z, x, and y, but its solution in the general case is so difficult, that nothing will be gained by attempting it.

In the particular case, however, where z is a function of x only, its value can be determined, as we shall then have

$$\frac{dz}{dy} = 0,$$

and equation (1) will reduce to

$$-\frac{Qdz}{dx} + \left(\frac{dP}{dy} - \frac{dQ}{dx}\right)z = 0,$$

or

$$\frac{dz}{z} = \frac{1}{Q}\left(\frac{dP}{dy} - \frac{dQ}{dx}\right)dx.$$

But by hypothesis z is a function of x, therefore

$$\frac{1}{Q}\left(\frac{dP}{dy} - \frac{dQ}{dx}\right) = f(x) = X;$$

then

$$\int \frac{dz}{z} = \int Xdx;$$

whence

$$lz = \int Xdx, \qquad z = e^{\int Xdx}$$

Let this be illustrated by the example

$$dx + 2xydy + 2y^2dx = 0,$$

in which

$$P = 1 + 2y^2 \qquad Q = 2xy;$$

whence

$$\frac{1}{Q}\left(\frac{dP}{dy} - \frac{dQ}{dx}\right) = \frac{1}{x} = X,$$

and

$$z = e^{\int X dx} = e^{\int \frac{1}{x} dx} = e^{lx} = x,$$

x being the common factor, the immediate differential equation must be

$$xdx + 2x^2ydy + 2xy^2dx = 0,$$

which can be integrated as in article (168).

In a similar way, if $z = f'(y)$, its value may be determined.

179. Differential equations of the first order, containing the higher powers of dy, may arise, as in the last case of article (56), from the elimination of the higher powers of a constant. Such equations, after division by dx^n, may be put under the form

$$\left(\frac{dy}{dx}\right)^n + U'\left(\frac{dy}{dx}\right)^{n-1} \ldots\ldots\ldots\ldots U^{n'} = 0 \ldots\ldots\ldots\ldots (1).$$

The determination of the primitive equation will then depend upon the solution of equation (1), or upon the division of the first member into its factors of the first degree. There are n such factors, and it is plain that each, when placed equal to zero and integrated, will give an equation between y and x which may be regarded as a primitive equation.

If, then, the values of $\frac{dy}{dx}$ be denoted by V, V′, V″, &c., equation (1) may be written

$$\left(\frac{dy}{dx} - V\right)\left(\frac{dy}{dx} - V'\right)\left(\frac{dy}{dx} - V''\right)\&c., = 0,$$

which may be satisfied by placing

$$\frac{dy}{dx} - V = 0, \qquad \frac{dy}{dx} - V' = 0, \&c. \ldots\ldots\ldots\ldots (2);$$

and if the integrals of these equations be denoted by P, P′, P″, &c., respectively, we shall have

$$PP'P''\&c. = 0 \ldots\ldots\ldots\ldots (3),$$

for the most general primitive equation, particular cases of which may be obtained by placing $P = 0$, $P' = 0$, or the product of any of these factors taken two and two, or three and three, &c.

It would appear, since a constant is to be added in the integration of each of the equations (2), that (3) ought to contain n arbitrary constants; but equation (1) can only be deduced from its primitive equation by the elimination of the nth power of a constant: [Or by raising $(\frac{dy}{dx} - V)$ to the nth power, in which case the primitive equation must be $y = \int V dx$]. It is plain then that the constants added ought to be equal, or that the same should be added in each integration.

The n differential equations of the first degree which are factors of (1) are readily accounted for, by supposing the primitive equation to be solved with reference to C, which will have n values, each of which differentiated will give one of the equations referred to.

As there will be difficulty in the solution of equation (1), when the degree is higher than the second, it will be well to discuss some particular cases which admit of integration by other means.

180. I. If the proposed equation does not contain y, and it be easier to solve it with reference to x than with reference to $\frac{dy}{dx}$, which we will denote by p, we can then obtain

$$x = f(p) \ldots\ldots\ldots\ldots (1).$$

But

$$dy = pdx,$$

and by parts, article (140),

$$y = px - \int xdp = px - \int f(p)dp + C;$$

whence, if $f(p)dp$ can be integrated, p may be eliminated by the aid of equation (1), and the primitive equation between x, y and C, deduced.

II. If the proposed equation does not contain x, and may be solved with reference to y, we shall have

$$y = f(p) \ldots\ldots\ldots\ldots (3),$$

$$dy = d\,f(p) \qquad \text{or} \qquad pdx = d\,f(p);$$

whence

$$dx = \frac{df(p)}{p}, \qquad x = \int \frac{df(p)}{p} + C.$$

Combining this with equation (3), and eliminating p, a primitive equation will result between x, y and C.

III. When both variables enter, but y enters only to the first power, we may take its value in terms of p and x, differentiate it, and thus obtain

$$dy = \mathrm{R}dx + \mathrm{S}dp;$$

or, since $dy = pdx$,

$$(\mathrm{R} - p)dx + \mathrm{S}dp = 0.$$

If this can be integrated, the result may be combined with the proposed equation, p eliminated, and a primitive equation between y and x determined.

Suppose the deduced value of y to be

$$y = px + \mathrm{P} \ldots\ldots\ldots\ldots (4),$$

in which $\mathrm{P} = f(p)$. By differentiation, we obtain

$$dy = pdx + xdp + \frac{d\mathrm{P}}{dp}dp;$$

or

$$(x + \frac{d\mathrm{P}}{dp})dp = 0,$$

which may be satisfied by making

$$x + \frac{d\mathrm{P}}{dp} = 0 \ldots\ldots (5), \qquad \text{or} \qquad dp = 0 \ldots\ldots (6).$$

Equation (6) gives $p = \mathrm{C}$;

whence by substitution in (4),

$$y = Cx + C',$$

C′ being what P becomes when $p = C$.

Equation (5) expresses a relation between x and p, and if it be combined with (4), and p be eliminated, an equation between x and y will result, which will contain no arbitrary constant.

Let there be for a particular example

$$ydx - xdy = n\sqrt{dx^2 + dy^2};$$

whence

$$y = px + n\sqrt{1 + p^2} \ldots\ldots (7),$$

$$dy = pdx + xdp + \frac{npdp}{\sqrt{1 + p^2}},$$

$$dp\left(x + \frac{np}{\sqrt{1 + p^2}}\right) = 0;$$

whence

$$x + \frac{np}{\sqrt{1 + p^2}} = 0, \quad dp = 0 \quad \text{or} \quad p = C.$$

This value of p in (7) gives

$$y = Cx + n\sqrt{1 + C^2}.$$

From the other factor we have

$$p = \pm \frac{x}{\sqrt{n^2 - x^2}},$$

which in (7), gives

$$y^2 + x^2 = n^2,$$

a result containing no arbitrary constant, which will be further explained in the following article.

SINGULAR SOLUTIONS.

181. It has been seen, that many differential equations of the first order result from the elimination of a constant, from the primitive equation and its immediate differential. Thus, let

$$f(x, y, C) = 0 \ldots\ldots (1),$$

be the primitive equation containing the variables x and y, and the constant C,

$$Pdx + Qdy = 0 \ldots\ldots (2)$$

its immediate differential equation, and

$$P'dx + Q'dy = 0 \ldots\ldots (3),$$

the result obtained by the elimination of C from (1) and (2). It may now be asked; may not such a function of x and y be substituted for C, that the result of the combination of equation (1) under this supposition, and its immediate differential, shall be the same as before? To answer this, let equation (1) be differentiated, x, y, and C being regarded as variables, we thus obtain

$$Pdx + Qdy + C'dC = 0 \ldots\ldots (4).$$

Now if $C'dC = 0$, it is plain that equation (4) will be the same as equation (2), and the result of the elimination of C between it and (1), will then be the same as equation (3).

If then for C in equation (1), we substitute the variable value deduced from the equation

$$C'dC = 0,$$

that equation will contain no arbitrary constant, and yet will be as much a primitive equation, as any one containing the arbitrary constant.

Such results are termed *singular solutions*, inasmuch as they can not possibly be obtained from the complete integral, Art. (132), by assigning to the arbitrary constant a particular value; the latter results being called *particular integrals.*

The equation $C'dC = 0$ can be satisfied, by making

$$dC = 0 \qquad \text{or} \qquad C' = 0.$$

The first gives $C =$ a constant, the particular values of which when substituted in equation (1) give particular integrals.

The values of C deduced from $C' = 0$, if variable, will then give the only singular solutions.

To illustrate, let us resume the complete integral of equation (7), in the preceding article,

$$y = Cx + n\sqrt{1 + C^2} \ldots\ldots (5).$$

Differentiating with reference to C, we have

$$0 = xdC + \frac{nCdC}{\sqrt{1 + C^2}};$$

whence

$$x + \frac{nC}{\sqrt{1+C^2}} = 0\ldots\ldots(6),$$

and

$$x^2 + x^2C^2 = n^2C^2, \qquad \text{or} \qquad C = -\frac{x}{\sqrt{n^2 - x^2}};$$

the negative value of C being plainly the only one which will satisfy equation (6). Its substitution in (5), gives

$$y = -\frac{x^2}{\sqrt{n^2 - x^2}} + n\sqrt{\frac{n^2}{n^2 - x^2}},$$

$$y = \sqrt{n^2 - x^2} \qquad \text{or} \qquad y^2 + x^2 = n^2,$$

the singular solution found in the preceding article.

INTEGRATION OF DIFFERENTIAL EQUATIONS OF THE SECOND ORDER.

182. Of these equations, which in their most general form contain $\frac{d^2y}{dx^2}$, $\frac{dy}{dx}$, y, x, and constants, we shall only discuss those particular cases which admit of integration.

I. The proposed equation may contain only $\frac{d^2y}{dx^2}$, x, and constants; in which case, solving it with reference to $\frac{d^2y}{dx^2}$, we have

$$\frac{d^2y}{dx^2} = f(x),$$

which may be integrated as in article (164).

183. II. It may contain only $\frac{d^2y}{dx^2}$, y, and constants. Solving the equation as before, we obtain

$$\frac{d^2y}{dx^2} = \mathrm{Y}.$$

Multiplying by $2dy$,

$$\frac{2dy}{dx}\frac{d^2y}{dx} = 2\mathrm{Y}dy,$$

and integrating,

$$\frac{dy^2}{dx^2} = \int 2\mathrm{Y}dy + \mathrm{C}, \qquad \text{or} \qquad \frac{dy}{dx} = \sqrt{2\int \mathrm{Y}dy + \mathrm{C}};$$

whence

$$dx = \frac{dy}{\sqrt{2\int \mathrm{Y}dy + \mathrm{C}}}, \qquad x = \int \frac{dy}{\sqrt{2\int \mathrm{Y}dy + \mathrm{C}}} + \mathrm{C}'$$

Examples.

1. If $$a^2d^2y + ydx^2 = 0,$$

$$\frac{d^2y}{dx^2} = -\frac{y}{a^2} \qquad \frac{2dy}{dx}\frac{d^2y}{dx} = -\frac{2ydy}{a^2},$$

$$\frac{dy^2}{dx^2} = -\frac{y^2}{a^2} + \mathrm{C}, \qquad \frac{dy}{dx} = \sqrt{\mathrm{C} - \frac{y^2}{a^2}},$$

$$x = \int \frac{dy}{\sqrt{C - \frac{y^2}{a^2}}} + C',$$

which may be integrated as in case I, article (133).

2. Let $$d^2y\sqrt{ay} = dx^2.$$

184. III. The equation may contain only $\frac{d^2y}{dx^2}$, $\frac{dy}{dx}$, and constants, being expressed generally thus,

$$F\left(\frac{d^2y}{dx^2}, \frac{dy}{dx}\right) = 0 \ldots\ldots\ldots\ldots(1).$$

Make $\frac{dy}{dx} = p$; then $\frac{d^2y}{dx^2} = \frac{dp}{dx}$, and (1) becomes

$$F\left(\frac{dp}{dx}, p\right) = 0,$$

which is of the first order with reference to dp, and may be solved with reference to dx; whence

$$dx = F'(p)dp \ldots\ldots(2), \qquad x = \int F'(p)dp + C \ldots\ldots(3).$$

Multiplying (2) by p, we have

$$pdx = dy = pF'(p)dp; \qquad y = \int pF'(p)dp + C' \ldots\ldots(4).$$

Eliminating p from (3) and (4), we have the primitive equation between x, y, and the two arbitrary constants C and C'.

For an example, let

$$\frac{(dx^2 + dy^2)^{\frac{3}{2}}}{dxd^2y} = a, \qquad \text{or} \qquad \frac{dx(1 + p^2)^{\frac{3}{2}}}{dp} = a;$$

whence

$$dx = \frac{adp}{(1+p^2)^{\frac{3}{2}}}, \qquad pdx = dy = \frac{apdp}{(1+p^2)^{\frac{3}{2}}}.$$

Integrating the last two expressions, we have

$$x = C + \frac{ap}{\sqrt{1+p^2}}, \qquad y = C' - \frac{a}{\sqrt{1+p^2}},$$

and eliminating p,

$$(x - C)^2 + (y - C')^2 = a^2,$$

as was to be expected, since the proposed equation expresses a constant radius of curvature.

185. IV. If the given equation does not contain y, it may be expressed

$$F\left(\frac{d^2y}{dx^2}, \frac{dy}{dx}, x\right) = 0, \qquad \text{or} \qquad F\left(\frac{dp}{dx}, p, x\right) = 0,$$

which is of the first order with reference to dp. Its integral will give an equation of the form

$$f(p, x) = 0,$$

in which, p being replaced by $\frac{dy}{dx}$, and the result integrated, we shall have

$$f'(y, x) = 0,$$

with two arbitrary constants.

For an example, let

$$\frac{d^2y}{dx^2} = \frac{dy}{dx}\frac{1}{x},$$

or

$$\frac{dp}{dx} = \frac{p}{x}, \qquad \frac{dp}{p} = \frac{dx}{x},$$

$$lp = lx + \mathrm{C}, \qquad p = \mathrm{C}'x,$$

$$\frac{dy}{dx} = \mathrm{C}'x, \quad \text{and} \quad y = \frac{\mathrm{C}'x^2}{2} + \mathrm{C}''.$$

186. V. If the given equation does not contain x, it may be expressed

$$\mathrm{F}\left(\frac{d^2y}{dx^2}, \frac{dy}{dx}, y\right) = 0 \text{............}(1).$$

Since $dy = pdx$,

$$dx = \frac{dy}{p}, \qquad \frac{d^2y}{dx^2} = \frac{dp}{dx} = \frac{pdp}{dy},$$

and equation (1) may be written

$$\mathrm{F}\left(\frac{pdp}{dy}, p, y\right) = 0,$$

which is of the first order with reference to dp and dy. Its integral will then be expressed

$$\mathrm{F}'(p, y) = 0, \qquad \text{or} \qquad \mathrm{F}'\left(\frac{dy}{dx}, y\right) = 0,$$

and this may be treated as in case II. Art. (180).

187. VI. If the equation be of the form,

$$\frac{d^2y}{dx^2} + X\frac{dy}{dx} + X'y = 0\text{............}(1).$$

Make $$y = e^{\int udx}\text{............}(2);$$

then

$$\frac{dy}{dx} = ue^{\int udx}, \qquad \frac{d^2y}{dx^2} = e^{\int udx}\left(u^2 + \frac{du}{dx}\right).$$

These values in (1) give, (since the common factor $e^{\int udx}$ disappears,)

$$\frac{du}{dx} + u^2 + Xu + X' = 0,$$

which is of the first order with reference to du. After integration, the value of u being determined and substituted in (2), will give the required primitive equation,

$$y = e^{\int fxdx}.$$

INTEGRATION OF DIFFERENTIAL EQUATIONS OF HIGHER ORDERS THAN THE SECOND.

188. Of these, it will also be sufficient for our purpose to discuss a few of the most simple cases.

I. Suppose the equation to contain only $\frac{d^ny}{dx^n}$, $\frac{d^{n-1}y}{dx^{n-1}}$, and constants, it may then be expressed,

$$F\left(\frac{d^ny}{dx^n}, \frac{d^{n-1}y}{dx^{n-1}}\right) = 0 \ldots\ldots\ldots\ldots(1).$$

Make

$$\frac{d^{n-1}y}{dx^{n-1}} = u; \qquad \text{then} \qquad \frac{d^ny}{dx^n} = \frac{du}{dx}.$$

These values in (1) give

$$F\left(\frac{du}{dx}, u\right) = 0,$$

which is of the first order, and its integral will give u in terms of x, or

$$u = X + C, \qquad \frac{d^{n-1}y}{dx^{n-1}} = X + C,$$

and finally,

$$y = \int^{n-1}(X + C)dx^{n-1}.$$

189. II. Suppose the equation expressed thus,

$$F\left(\frac{d^ny}{dx^n}, \frac{d^{n-2}y}{dx^{n-2}}\right) = 0 \ldots\ldots\ldots\ldots(1).$$

Make

$$\frac{d^{n-2}y}{dx^{n-2}} = u, \qquad \text{then} \qquad \frac{d^ny}{dx^n} = \frac{d^2u}{dx^2},$$

and equation (1) will become

$$F\left(\frac{d^2u}{dx^2}, u\right) = 0,$$

which may be integrated as in article (183), and the value of $u = f(x)$ determined; we shall then have

$$\frac{d^{n-2}y}{dx^{n-2}} = f(x) \qquad \text{and} \qquad y = \int^{n-2} f(x)dx^{n-2}.$$

190. III. Suppose the equation to be of the form

$$d^3y + Ad^2ydx + Bdydx^2 + Dydx^3 = 0\text{.........}(1).$$

Make

$$y = e^u\text{..........}(2),$$

u being an arbitrary function of x; then

$$dy = e^u du \qquad d^2y = e^u(d^2u + du^2)$$

$$d^3y = e^u(d^3u + 3dud^2u + du^3).$$

These values in (1) give

$$d^3u + 3dud^2u + du^3 + A(d^2u + du^2)dx$$

$$+ Bdudx^2 + Ddx^3 = 0\text{ ....}(3).$$

Since u in equation (2) is arbitrary, let such a value be assigned to it, that its differential shall be constant, in which case

$$du = mdx, \qquad d^2u = 0, \qquad d^3u = 0.$$

Equation (3), under this supposition, reduces to

$$m^3 + Am^2 + Bm + D = 0\text{..........}(4).$$

From this equation we may determine the value of the constant m. Denoting the three roots by

$$m, \quad m', \quad m'',$$

we have for du the three values

$$du = mdx, \qquad du = m'dx, \qquad du = m''dx;$$

whence

$$u = mx + C, \qquad u = m'x + C', \qquad u = m''x + C'',$$

and

$$y = e^{mx+C}, \qquad y = e^{m'x+C'}, \qquad y = e^{m''x+C''};$$

or calling

$$e^{C} = C, \qquad e^{C'} = C', \qquad e^{C''} = C'',$$

$$y = Ce^{mx}, \qquad y = C'e^{m'x}, \qquad y = C''e^{m''x}.$$

But since these values of y each contain but one arbitrary constant, they must be particular cases of the general value of y, which must be of such a form that either of the above particular values can be deduced from it; that is,

$$y = Ce^{mx} + C'e^{m'x} + C''e^{m''x},$$

from which the first particular value is deduced by making C' and $C'' = 0$; and in a similar way, the others.

If two of the roots m, m', m'', are equal, that is, if $m = m'$, we should have the equation

$$y = (C + C')e^{mx} + C''e^{m''x} = Ce^{mx} + C''e^{m''x},$$

containing but *two* arbitrary constants, $C + C'$ being denoted by C. It is not then general. But in this case, $y = Ce^{mx}$ being a particular value,

$$y = C'xe^{mx} \ldots\ldots\ldots\ldots (5)$$

will be another; for, differentiating it, we have

$$dy = C'e^{mx}(1 + mx)dx,$$

$$d^2y = C'e^{mx}(2m + m^2x)dx^2,$$

$$d^3y = C'e^{mx}(3m^2 + m^3x)dx^3,$$

and these substituted in equation (1), give

$$(m^3 + Am^2 + Bm + D)x + (3m^2 + 2Am + B) = 0 \ldots\ldots\ldots (6).$$

But the coefficient of x is the same as the first member of equation (4), which has two roots equal to m; and $3m^2 + 2Am + B$ is its first derived polynomial, which, when placed equal to 0, must have one root equal to m (see Algebra); hence both terms of (6) are 0, and $y = C'xe^{mx}$ satisfies the given differential equation, and must therefore be a particular value of the general one,

$$y = Ce^{mx} + C'xe^{mx} + C''e^{m''x}.$$

If $m = m' = m''$, it may be shown also by trial, as above, that

$$y = C''x^2e^{mx}$$

is a particular value; whence the general value must be

$$y = e^{mx}(C + C'x + C''x^2).$$

Two of the roots may be imaginary, but as the discussion in this case is quite complicated, and of little value to the student, we omit it.

To illustrate the above, let

$$d^3y + 2d^2ydx - dydx^2 - 2ydx^3 = 0.$$

Comparing this with equation (1), we have

$$A = 2, \qquad B = -1, \qquad D = -2;$$

and equation (4) becomes

$$m^3 + 2m^2 - m - 2 = 0;$$

whence

$$m = -2, \qquad 1, \qquad \text{and} \quad -1,$$

and the general value of y is

$$y = Ce^{-2x} + C'e^{x} + C''e^{-x}.$$

191. It is plain that the preceding principles can readily be extended to the general equation

$$d^ny + Ad^{n-1}ydx + Bd^{n-2}ydx^2 \ldots\ldots + Mydx^n = 0,$$

and that the general value of y will be

$$y = Ce^{mx} + C'e^{m'x} + C''e^{m''x} + \&c. \ldots\ldots$$

192. If the equation be

$$d^3y + Xd^2ydx + X'dydx^2 + X''ydx^3 = 0 \ldots\ldots (1),$$

in which X, &c. are functions of x, the difficulty of integration is much increased. If, however, we know three particular values of

y, Cy', $C'y''$, $C''y'''$, each of which will satisfy the given equation, then the general value of y will equal their sum, that is

$$y = Cy' + C'y'' + C''y'''......(2).$$

To verify this, let equation (2) be differentiated three times and the proper values substituted in (1), we shall thus obtain

$$\left.\begin{array}{l} C(d^3y' + Xd^2y'dx + X'dy'dx^2 + X''y'dx^3) \\ + C'(d^3y'' + Xd^2y''dx + X'dy''dx^2 + X''y''dx^3) \\ + C''(d^3y''' + Xd^2y'''dx + X'dy'''dx^2 + X''y'''dx^3) \end{array}\right\} = 0,$$

which is satisfied, since each of the three terms is by hypothesis equal to 0.

193. The above demonstration can be generalized, and a similar result obtained for the equation

$$d^ny + Xd^{n-1}ydx + ...yX^{(n-1)'}dx^n = 0.$$

This, and the equations discussed in the three preceding articles, belong to the class termed *linear*. See note to article (176).

INTEGRATION OF PARTIAL DIFFERENTIAL EQUATIONS OF THE FIRST ORDER.

194. A partial differential equation of the first order, derived from an equation between the three variables z, y and x, z being regarded as a function of x and y, contains in its most general form, the three variables, the two partial differential coefficients,

$\frac{dz}{dx}$ and $\frac{dz}{dy}$, and constants. Without attempting to discuss the most general, we will confine ourselves to a few of the most simple cases.

I. If the equation contains but one partial differential coefficient, and the two independent variables, that is, if

$$\frac{dz}{dx} = \mathrm{P},$$

P being a function of x and y; we integrate at once as in article (166). For example, if

$$\frac{dz}{dx} = \frac{x}{\sqrt{x^2 + y^2}}, \qquad z = \sqrt{x^2 + y^2} + \mathrm{Y}.$$

195. II. Let the equation be

$$\frac{dz}{dx} = \mathrm{R},$$

R being a function of the three variables. Since the partial differential coefficient has been obtained under the supposition that y is constant, the proposed equation may be regarded as a differential equation between z and x, and may be integrated as in article (172), taking care to add an arbitrary function of y.

Examples.

1. Let $$\frac{dz}{dx} = \frac{\sqrt{y^2 - z^2}}{z}\,x.$$

By the separation of the variables, we have

$$xdx = \frac{zdz}{\sqrt{y^2 - z^2}},$$

and by integration

$$\frac{x^2}{2} = -\sqrt{y^2 - z^2} + \varphi(y)$$

2. Let
$$\frac{dz}{dx} = \frac{y^2 + z^2}{y^2 + x^2}.$$

196. III. Let the equation be

$$M\frac{dz}{dy} + N\frac{dz}{dx} = 0,$$

M and N being functions of x and y.

Solving the equation with reference to $\frac{dz}{dy}$, we have

$$\frac{dz}{dy} = -\frac{N}{M}\frac{dz}{dx}.$$

But since z is a function of x and y,

$$dz = \frac{dz}{dx}dx + \frac{dz}{dy}dy,$$

or by the substitution of the value of $\frac{dz}{dy}$,

$$dz = \frac{dz}{dx}(dx - \frac{N}{M}dy) = \frac{dz}{dx}\left(\frac{Mdx - Ndy}{M}\right) \ldots\ldots(1).$$

If S be the factor which will make $Mdx - Ndy$ integrable, we may write

$$S(Mdx - Ndy) = du,$$

which in (1), gives

$$dz = \frac{1}{SM}\frac{dz}{dx}du,$$

to satisfy which, it is only necessary that $\frac{1}{SM}\frac{dz}{dx} = F(u)$; whence

$$dz = F(u)du \qquad z = \varphi(u),$$

the form of this function being arbitrary.

Examples.

1\. If
$$x\frac{dz}{dy} - y\frac{dz}{dx} = 0,$$

$$Mdx - Ndy = xdx + ydy,$$

which is made integrable by the factor 2, and we have

$$x^2 + y^2 = u, \qquad \text{and} \qquad z = \varphi(x^2 + y^2),$$

which is the general equation of a surface of revolution.

2\. If
$$y\frac{dz}{dy} + x\frac{dz}{dx} = 0,$$

$$Mdx - Ndy = ydx - xdy,$$

which may be integrated by the aid of the factor $\frac{1}{y^2}$; whence

$$\frac{x}{y} = u, \qquad \text{and} \qquad z = \varphi\left(\frac{x}{y}\right).$$

RECTIFICATION OF CURVES.

197. The operation by which the length of a curve is determined, is called its *rectification*; and when a right line can be constructed equal in length to a definite portion of it, the curve is said to be *rectifiable.*

If x and y are the co-ordinates of any point of a curve z, we have, article (86),

$$dz = \sqrt{dx^2 + dy^2}, \qquad \text{or} \qquad z = \int\sqrt{dx^2 + dy^2},$$

which is a general expression for the length of an indefinite portion of any curve.

In order then to obtain the length of a particular curve; *we differentiate its equation, and by combining the result with the given equation, deduce the value of dy in terms of x and dx, or of dx in terms of y and dy, and substitute in the general expression for the differential of an arc. This will then contain but one variable and its differential, and the integral will express the length of an indefinite portion of the curve.*

If the length of a definite portion be required, the integral must be taken between the limits, designated by the two values of the variable belonging to the extremities of this definite portion, Art. (132). If the expression thus obtained can be constructed geometrically, the curve is rectifiable.

198. Let these principles be applied to the rectification of the parabolas given by the general equation

$$y^n = px^m.$$

This can be written

$$y = p^{\frac{1}{n}}x^{\frac{m}{n}} = p'x^r\ldots\ldots(1).$$

By differentiation, we have

$$dy = rp'x^{r-1}dx\,;$$

hence

$$z = \int\sqrt{dx^2 + dy^2} = \int dx(1 + r^2p'^2x^{2r-2})^{\frac{1}{2}}.$$

This admits of an integral, in a finite number of algebraic terms and may be constructed, when either

$$\frac{1}{2r-2}, \qquad \text{or} \qquad -\left(\frac{1}{2r-2} + \frac{1}{2}\right),$$

is equal to a positive whole number, Art. (147).

Denoting these numbers by k and k', we have

$$\frac{1}{2r-2} = k, \qquad -\left(\frac{1}{2r-2} + \frac{1}{2}\right) = k'\,;$$

whence

$$r = \frac{2k+1}{2k}\,; \qquad \text{and} \qquad r = \frac{2k'}{2k'+1}.$$

These values of r in equation (1), give

$$y = p'x^{\frac{2k+1}{2k}}, \qquad \text{and} \qquad y = p'x^{\frac{2k'}{2k'+1}} \ldots\ldots(2).$$

Whenever the equation is a particular case of either of these forms, the parabola represented by it is rectifiable.

As the second of equations (2) will become of the same form as the first, by changing x into y and y into x, they will represent curves of the same kind, and all the cases of rectification may therefore be deduced by the discussion of either.

If in the first, we make $k = 1$, we have

$$y = p'x^{\frac{3}{2}}, \qquad \text{or} \qquad y^2 = p'^2x^3,$$

which is the equation of a cubic parabola. In this case $r = \dfrac{3}{2}$ and

$$z = \int dx(1 + \frac{9}{4}p'^2x)^{\frac{1}{2}} = \frac{8}{27p'^2}(1 + \frac{9}{4}p'^2x)^{\frac{3}{2}} + C.$$

If we wish the length from that point whose abscissa is a, to that whose abscissa is b, we take the integral between the limits a and b.

Let us, however, estimate the arc from the vertex, or suppose the origin of the integral to be $x = 0$, Art. (132); we then have

$$0 = \frac{8}{27p'^2} + C, \qquad \text{or} \qquad C = -\frac{8}{27p'^2};$$

whence, denoting this particular integral by z',

$$z' = \frac{8}{27p'^2}[(1 + \frac{9}{4}p'^2x)^{\frac{3}{2}} - 1],$$

for the length of any arc whose abscissa is x.

199. By differentiating the equation

$$y^2 = 2px,$$

we obtain

$$2ydy = 2pdx, \qquad dx = \frac{ydy}{p}.$$

This value in the expression $z = \int \sqrt{dx^2 + dy^2}$, gives

$$z = \int dy \sqrt{1 + \frac{y^2}{p^2}} = \frac{1}{p} \int dy (p^2 + y^2)^{\frac{1}{2}},$$

which by formula **B** may be reduced to

$$z = \frac{y\sqrt{p^2 + y^2}}{2p} + \frac{p}{2} \int \frac{dy}{\sqrt{p^2 + y^2}}.$$

But

$$\int \frac{dy}{\sqrt{p^2 + y^2}} = l(\sqrt{p^2 + y^2} + y) + C \ldots\ldots \text{Art. } (144);$$

hence

$$z = \frac{y\sqrt{p^2 + y^2}}{2p} + \frac{p}{2} l(\sqrt{p^2 + y^2} + y) + C.$$

If we estimate the arc from the vertex, where $y = 0$, we have

$$0 = \frac{p}{2} lp + C, \qquad \text{or} \qquad C = -\frac{p}{2} lp,$$

and finally, denoting the particular integral by z',

$$z' = \frac{y\sqrt{p^2 + y^2}}{2p} + \frac{p}{2}\,[l(\sqrt{p^2 + y^2} + y) - lp],$$

which is transcendental and can not be constructed. Hence the common parabola is not rectifiable.

200. For the arc of the circle, we have, Art. (86),

$$z = \mathrm{R}\int \frac{dx}{\sqrt{\mathrm{R}^2 - x^2}},$$

which *can only be expressed by a series* and therefore admits of no construction.

Differentiating the equation of the ellipse, we deduce

$$dy = -\,\frac{b^2x}{a^2y}\,dx\,;$$

whence

$$z = \int dx\sqrt{1 + \frac{b^4x^2}{a^4y^2}} = \frac{1}{a}\int \frac{dx\sqrt{a^4 - (a^2 - b^2)x^2}}{\sqrt{a^2 - x^2}},$$

which can only be expressed by a series.

201. The differential equation of the cycloid, Art. **(114)**, is

$$dx = \frac{ydy}{\sqrt{2ry - y^2}}.$$

By the substitution of this value of dx, we obtain

$$z = \int \sqrt{dx^2 + dy^2} = \int dy\sqrt{\frac{2ry}{2ry - y^2}} = \sqrt{2r}\int dy(2r - y)^{-\frac{1}{2}}\,;$$

whence, article (129),

$$z = -2\sqrt{2r}(2r - y)^{\frac{1}{2}} + C = -2\sqrt{2r(2r - y)} + C.$$

If we estimate the arc from the point D, where $y = 2r$, we have

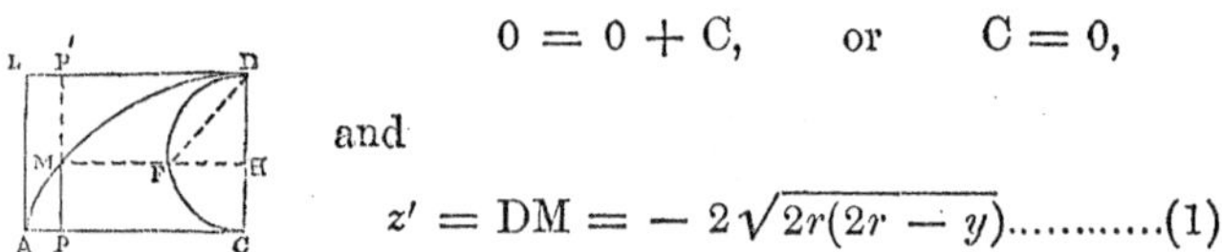

$$0 = 0 + C, \qquad \text{or} \qquad C = 0,$$

and

$$z' = DM = -2\sqrt{2r(2r - y)} \text{..........} (1).$$

From the figure we see that

$$DF = \sqrt{DC \times DH} = \sqrt{2r(2r - y)},$$

hence

$$DM = -2DF,$$

or the arc is rectifiable and equal to *twice the corresponding chord of the generating circle.*

If in equation (1) we make $y = 0$, and denote the definite integral by z'', we have

$$z'' = DMA = -4r = -2DC,$$

as in article (118).

202. For the rectification of the spirals we take the expression in article (120),

$$dz = \sqrt{du^2 + u^2dt^2}.$$

By differentiating the general equation $u = at^n$, we deduce

$$du^2 = n^2a^2t^{2n-2}dt^2;$$

whence by substitution, &c.,

$$z = \int at^{n-1}dt\sqrt{n^2 + t^2}.$$

For the logarithmic spiral, when M = 1, we have

$$t = lu, \qquad dt = \frac{du}{u},$$

$$z = \int du\sqrt{2} = u\sqrt{2} + C;$$

or, estimating the arc from the pole, where $u = 0$, we have

$$z' = u\sqrt{2},$$

or *the diagonal of the square upon the radius vector.*

QUADRATURE OF CURVES.

203. The quadrature of a curve is the operation by which the area included within it is determined; and a curve is *quadrable* when a square can be constructed equivalent to this area.

In article (88), we have

$$ds = ydx, \qquad \text{or} \qquad s = \int ydx \ldots\ldots(1),$$

in which s represents the indefinite area limited by the curve and the axis of X.

To obtain the value of s for any particular curve, *we take the value of y in terms of x from the equation of the curve, or the value of dx in terms of y and dy from its differential equation, and substitute in the formula* $s = \int y dx$; *the result obtained by integration will be the indefinite area.*

204. The value of y taken from the general equation of parabolas, Art. (198), is

$$y = p'x^r \ldots\ldots\ldots\ldots (1),$$

which, in the formula, gives

$$s = \int p'x^r dx = \frac{p'x^{r+1}}{r+1} + C.$$

If we estimate the area from the origin, where $x = 0$, we have

$$C = 0;$$

whence

$$s' = \frac{p'x^{r+1}}{r+1} = \frac{yx}{r+1},$$

that is, the area of a portion of a parabola, included between the curve, the axis of X, and any assumed ordinate, is equal to the rectangle of the ordinate and corresponding abscissa, divided by $r + 1$. Hence all parabolas are quadrable.

The same result may be obtained otherwise, thus: The value of x from (1) is

$$x = \frac{y^{\frac{1}{r}}}{p'^{\frac{1}{r}}}, \qquad \text{whence} \qquad dx = \frac{y^{\frac{1}{r}-1}\,dy}{rp'^{\frac{1}{r}}},$$

and this, in the formula, gives

$$s = \int \frac{y^{\frac{1}{r}}dy}{rp'^{\frac{1}{r}}} = \frac{1}{r+1} \cdot \frac{1}{p'^{\frac{1}{r}}}\, y^{\frac{1}{r}+1} = \frac{yx}{r+1} + \mathrm{C},$$

as before.

For the common parabola, we have $r = \frac{1}{2}$; whence

$$s' = \frac{yx}{\frac{1}{2}+1} = \frac{2}{3}\,xy.$$

For the cubic parabola, $r = \frac{3}{2}$; whence

$$s' = \frac{2}{5}\,xy.$$

205. The value of y taken from the equation of the ellipse referred to its centre and axes, is

$$y = \frac{b}{a}\sqrt{a^2 - x^2};$$

hence

$$s = \frac{b}{a}\int (a^2 - x^2)^{\frac{1}{2}}dx.$$

By formula $\mathfrak{B}_9$ we have

$$\int (a^2 - x^2)^{\frac{1}{2}}dx = \frac{1}{2}x(a^2 - x^2)^{\frac{1}{2}} + \frac{1}{2}a^2 \int dx\,(a^2 - x^2)^{-\frac{1}{2}}.$$

But

$$\int dx(a^2 - x^2)^{-\frac{1}{2}} = \int \frac{dx}{\sqrt{a^2 - x^2}} = \sin^{-1}\frac{x}{a} + C;$$

whence, finally,

$$s = \frac{b}{2a}x\sqrt{a^2 - x^2} + \frac{ab}{2}\sin^{-1}\frac{x}{a} + C;$$

which can not be constructed.

Taking the area between the limits $x = 0$ and $x = a$, we have

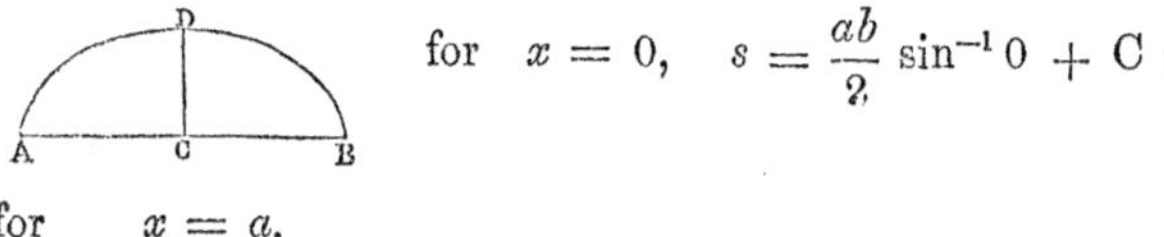

for $x = 0$, $s = \frac{ab}{2}\sin^{-1}0 + C = C;$

for $x = a$,

$$s = \frac{ab}{2}\sin^{-1}1 + C = \frac{ab}{2}\frac{\pi}{2} + C;$$

and for the difference, or definite integral,

$$s'' = \frac{ab}{4}\pi = CDB = \frac{1}{4}\text{th of the ellipse};$$

hence the entire area is πab.

If $a = b$, the ellipse becomes a circle of which a is the radius; whence the area of the circle is

$$\pi a^2 = \pi\,(\text{radius})^2.$$

The same result may be obtained by taking the value

$$y = \sqrt{2rx - x^2};$$

whence

$$s = \int dx\sqrt{2rx - x^2};$$

for the area of an indefinite portion of the circle.

206. In order to find an expression for the area of a portion of the hyperbola, it will be best to take its equation when referred to the centre and asymptotes,

$$xy = m,$$

and, since the asymptotes are oblique to each other, we must use the formula deduced in article (88),

$$ds = \sin\omega\, ydx,$$

ω being the angle included by the asymptotes.

The value $y = \dfrac{m}{x}$ being substituted in the formula, gives

$$ds = \sin\omega\,\frac{mdx}{x}; \qquad \text{whence} \qquad s = \sin\omega\, mlx + \mathrm{C}.$$

If we call the distance CB $= 1$, and estimate the area from the ordinate AB, for which $x = 1$, we have

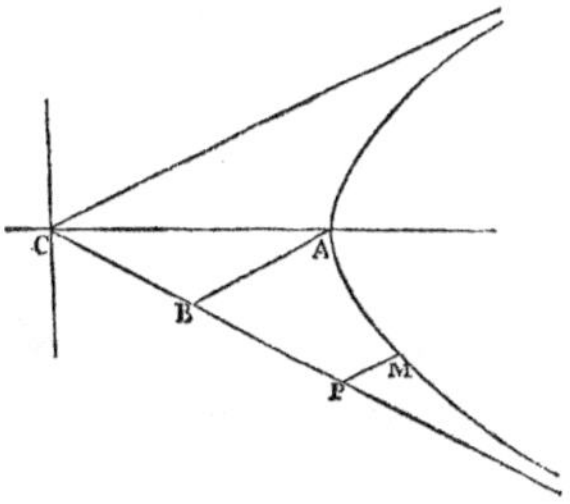

$$m = 1 \quad \text{and} \quad \mathrm{C} = 0;$$

whence

$$s' = \sin\omega\, lx;$$

or since sin ω may be regarded as the modulus of a new system of logarithms, we have

$$s' = \log x;$$

or, *the area between the curve and asymptote estimated from the ordinate of the vertex is equal to the logarithm of the abscissa of its extreme point, taken in a system whose modulus is the sine of the angle made by the asymptotes.*

207. The value of dx taken from the differential equation of the cycloid, and substituted in the expression $s = \int y dx$, gives

$$s = \int \frac{y^2 dy}{\sqrt{2ry - y^2}},$$

which can be reduced by formula $\mathfrak{B}_2$ and finally integrated.

A more simple method, however, is to obtain directly the area ALD. If we denote

P′M $= 2r - y$ by z, we shall have

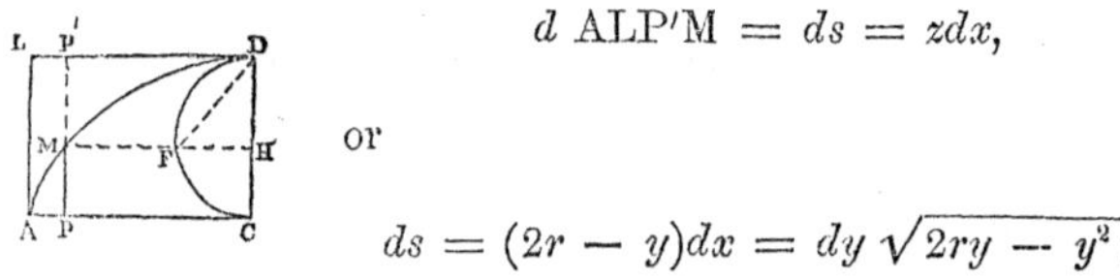

$$d\ \text{ALP'M} = ds = z dx,$$

or

$$ds = (2r - y)dx = dy\sqrt{2ry - y^2};$$

whence

$$s = \int dy \sqrt{2ry - y^2}.$$

But this is evidently the area of a portion of a circle whose radius is r, and abscissa y, Art. (205); that is, the area of the segment CFH. If we estimate these areas; the first from AL, and the second from the point C, they will both be 0, when $y = 0$; the arbitrary constant to be added in each case will then be 0, and we have

$$\text{ALP}'\text{M} = \text{CFH},$$

and when $y = 2r$,

$$\text{ALD} = \text{CFD} = \frac{\pi r^2}{2}.$$

But the area of the rectangle

$$\text{ALDC} = \text{AC} \times \text{CD} = \pi r.2r = 2\pi r^2;$$

hence

$$\text{area AMDC} = \text{ALDC} - \text{ALD} = \frac{3}{2}\pi r^2,$$

double of which, or the area included between one branch of the cycloid and its base, *is equal to three times the area of the generating circle.*

From this we see also, that the area, included between one branch of the cycloid and its base, is equal to *three fourths* of the rectangle described upon the base and axis, and this particular area would therefore appear to be quadrable; but in reality it is not so, for the base of this rectangle is the circumference of the given circle, to which it is not possible to construct a right line equal. In fact, y being the independent variable in the equation of the cycloid, Art. (114), it is impossible for any assumed value of y, to construct the corresponding abscissa, and thus the position of the points C, B, &c., can only be determined approximately.

208. For the logarithmic curve

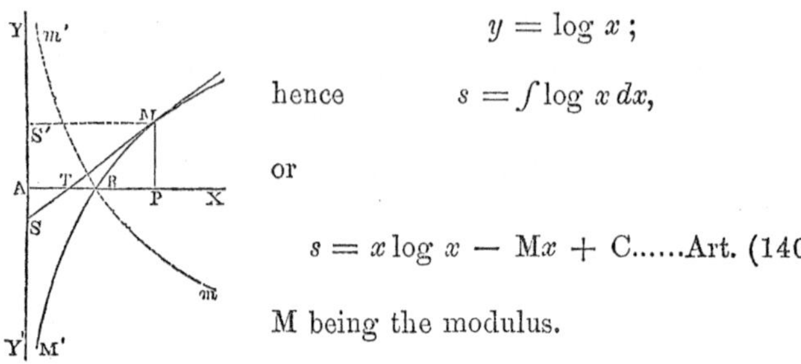

$$y = \log x\,;$$

hence $$s = \int \log x\, dx,$$

or

$$s = x \log x - \mathrm{M}x + \mathrm{C} \ldots\ldots \text{Art. (140)}.$$

M being the modulus.

If we estimate from the point B, where $x = 1$, we have

$$0 = -\mathrm{M} + \mathrm{C}, \qquad \mathrm{C} = \mathrm{M},$$

and

$$s' = x \log x - \mathrm{M}x + \mathrm{M}.$$

If we take the area included between the curve and axis of Y,

$$s = \int x dy = \int x\mathrm{M}\frac{dx}{x} = \mathrm{M}x + \mathrm{C},$$

or estimating from the line AB, for which $x = 1$,

$$\mathrm{C} = -\mathrm{M}\,; \qquad \text{whence} \qquad s' = \mathrm{M}(x - 1).$$

If $x = 0$, we have $s'' = -\mathrm{M} =$ area Y′ABM′.

If $x = 2$, " $s'' = \mathrm{M} =$ area ABMS′.

209. The curve given by the equation

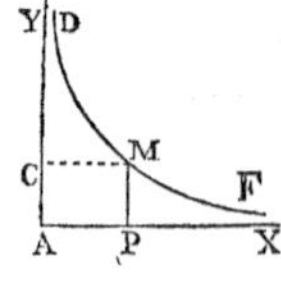

$$y^2 = \frac{1}{x},$$

to which, as in the figure, the axes of co-ordinates are asymptotes, presents a case worthy of notice.

By differentiation, we obtain

$$dx = -\frac{2dy}{y^3};$$

whence

$$s = -\int \frac{2dy}{y^2} = \frac{2}{y} + C.$$

Estimating the area from the line AY, where $y = \infty$, we have

$$0 = \frac{2}{\infty} + C, \qquad C = 0,$$

and

$$s' = \frac{2}{y}.$$

By making $\quad y = 1 = MP, \quad$ we have

$$s'' = 2 = APMD;$$

that is, the area APMD is finite and equal to twice the square APMC, although the curve does not touch the axis of Y at a finite distance.

If we take the area between the limits $y = 1$ and $y = 0$, we have

$$\text{area FMPX} = \frac{2}{0} - 2 = \infty.$$

210. For the quadrature of spirals, we take

$$ds = \frac{u^2dt}{2} \text{......Art. (120)}, \qquad \text{or} \qquad s = \int \frac{u^2dt}{2} \text{......(1)}.$$

The value of u^2 taken from the general equation of spirals, Art. (121), is $u^2 = a^2t^{2n}$. This substituted in formula (1), gives

$$s = \int \frac{a^2t^{2n}dt}{2} = \frac{a^2t^{2n+1}}{4n+2} + C.$$

Estimating the area from the pole, where $t = 0$ when n is positive, and ∞ when n is negative, we have, in all cases except when n is negative and numerically less than $\frac{1}{2}$, $C = 0$, and

$$s' = \frac{a^2t^{2n+1}}{4n+2}.$$

For the spiral of Archimedes, $n = 1$ and $a = \frac{1}{2\pi}$; whence

$$s' = \frac{t^3}{24\pi^2}.$$

If in this we make $t = 2\pi$, we have

$$s'' = \frac{\pi}{3},$$

which is the area PMA included within the first spire, or that described by one revolution of the radius vector. Since $PA = 1$, π represents the area of the circle PA; hence

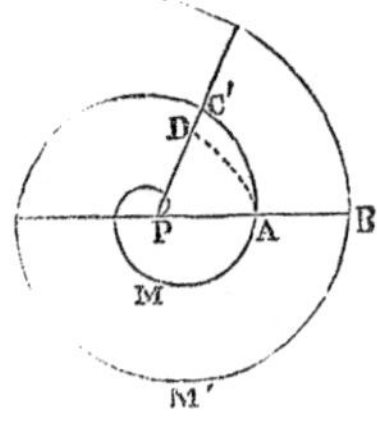

$$\text{area PMA} = \frac{1}{3} \text{ of the circle PA.}$$

If $t = 2\,(2\pi)$, we have

$$s'' = \frac{(4\pi)^3}{24\pi^2} = \frac{8}{3}\pi,$$

which is the whole area described by the radius vector during two revolutions. But it is plain that, during the second revolution, the part PMA will be described a second time ; hence, to obtain the area PAM′B, we must subtract that described during the first revolution ; we then have

$$\text{PAM}'\text{B} = \frac{8}{3}\pi - \frac{1}{3}\pi = \frac{7}{3}\pi ;$$

and in general it will be seen, that by each revolution of the radius vector, the area before described will be increased by the area from the pole out to the last spire ; hence, to obtain the area from the pole out to the mth spire, from the whole area described during m revolutions, take the area described during $m - 1$ revolutions, or take the integral between the limits
$t = (m - 1)2\pi$, and $t = m2\pi$, which gives

$$\frac{(m2\pi)^3}{24\pi^2} - \frac{[(m-1)2\pi]^3}{24\pi^2} = \frac{m^3 - (m-1)^3}{3}\pi .$$

The area terminated by the $(m + 1)$th spire is then

$$\frac{(m+1)^3 - m^3}{3}\pi ,$$

and the difference between the two expressions gives the area included between the mth and $(m + 1)$th spires, thus

$$\frac{(m+1)^3 - 2m^3 + (m-1)^3}{3}\pi = 2m\pi = m.2\pi .$$

If $m = 1$ in this expression, we have the area included between the first and second spire equal to 2π ; hence, in general, the area between the mth and $(m + 1)$th spires *is equal to m times that included between the first and second.*

If the area PAC$'$ be required, AC$'$ being a portion of the second spire corresponding to the arc $AD = \frac{2\pi}{n'}$, we should have for the whole area generated when the generating point has arrived at C$'$, since $t = 2\pi + \frac{2\pi}{n'}$,

$$s'' = \frac{\left(2\pi + \frac{2\pi}{n'}\right)^3}{24\pi^2},$$

from which, subtracting the area PMA, we have

$$APC' = \frac{\left(2\pi + \frac{2\pi}{n'}\right)^3}{24\pi^2} - \frac{(2\pi)^3}{24\pi^2} = \frac{\pi}{n'}\left(1 + \frac{1}{n'} + \frac{1}{3n'^2}\right),$$

or if we call AP (which has been regarded as unity) r,

$$APC' = \frac{\pi}{n'}\left(1 + \frac{1}{n'} + \frac{1}{3n'^2}\right)r^2.$$

If $AC' = \frac{1}{4}$ circumference $= \frac{2\pi}{4}$, then $n' = 4$, and

$$APC' = \frac{\pi}{4}\left(1 + \frac{1}{4} + \frac{1}{48}\right)r^2.$$

For the hyperbolic spiral $n = -1$, and the general value of s' becomes

$$s' = -\frac{a^2}{2t},$$

which is infinite when $t = 0$. For the integral between the limits $t = b$ and $t = c$, we have

$$s'' = \frac{a^2}{2}\left(\frac{1}{b} - \frac{1}{c}\right).$$

In the logarithmic spiral, when $M = 1$,

$$t = lu, \qquad dt = \frac{du}{u},$$

$$s = \int \frac{u^2dt}{2} = \int \frac{udu}{2} = \frac{u^2}{4} + C,$$

or estimating from the pole where $u = 0$ and $C = 0$; we have

$$s' = \frac{u^2}{4};$$

that is, *equal one-fourth the square described upon the radius vector of the extreme point of the curve.*

AREA OF SURFACES OF REVOLUTION.

211. In article (89), we have found for the differential of the area of a surface of revolution $du = 2\pi y\sqrt{dx^2 + dy^2}$; whence for the indefinite area, we have

$$u = \int 2\pi y \sqrt{dx^2 + dy^2} \ldots\ldots\ldots(1),$$

the axis of X being the axis of revolution, and $\sqrt{dx^2 + dy^2}$ the differential of the arc of the generating curve.

The indefinite area of any particular surface will then be obtained, *by deducing from the equation and differential equation of the meridian or generating curve, the values of* y *and* dy *in terms of* x *and* dx; *or of* dx *in terms of* y *and* dy, *and substituting in formula* (1). *The result of the integration will be the area required.*

212. Let the line AC, by its revolution about AB, generate the surface of a right cone. The origin of co-ordinates being at A, the equation of AC is

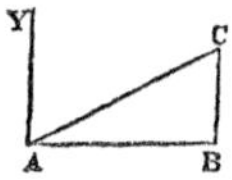

$$y = ax\,; \qquad \text{whence} \qquad dy = adx,$$

and

$$u = \int 2\pi axdx\sqrt{a^2+1} = \pi ax^2\sqrt{a^2+1} + \mathrm{C}.$$

Estimating the area from the vertex, where $x = 0$, we have $\mathrm{C} = 0$, and

$$u' = \pi ax^2\sqrt{a^2+1}.$$

Making $x = \mathrm{AB} = h$, we have the area of the cone whose altitude is h, and the radius of the base $\mathrm{BC} = b$,

$$u'' = \pi ah^2\sqrt{a^2+1},$$

or since $a = \dfrac{b}{h}$,

$$u'' = \frac{2\pi b\sqrt{b^2+h^2}}{2} = 2\pi b\frac{\mathrm{AC}}{2}\,;$$

that is, *the circumference of the base into half the side.*

213. From the equation of the circle, we have

$$y = \sqrt{2rx - x^2}, \qquad dy = \frac{(r-x)dx}{y}.$$

The surface of the sphere is then

$$u = \int 2\pi y \sqrt{dx^2 + \frac{(r-x)^2dx^2}{y^2}} = \int 2\pi r dx,$$

or

$$u = 2\pi rx + \mathrm{C}.$$

Taking the area between the limits $x = 0$, and $x = 2r$, we have

$$u'' = 4\pi r^2 = \textit{four great circles.}$$

214. From the equation of the ellipse, we have

$$y = \frac{b}{a}\sqrt{a^2 - x^2}, \qquad dy = -\frac{b^2x}{a^2y}dx;$$

whence for the area of the ellipsoid of revolution,

$$u = \int \frac{2\pi b}{a^2}dx\sqrt{a^4 - (a^2 - b^2)x^2}$$

$$= \frac{2\pi b}{a^2}\sqrt{a^2 - b^2}\int dx\sqrt{\frac{a^4}{a^2 - b^2} - x^2},$$

or placing $\frac{2\pi b}{a^2}\sqrt{a^2 - b^2} = \mathrm{C}'$, and $\frac{a^4}{a^2 - b^2} = \mathrm{R}'^2$,

$$u = \mathrm{C}'\int dx\sqrt{\mathrm{R}'^2 - x^2}.$$

But $\int dx\sqrt{\mathrm{R}'^2 - x^2}$ = area of a circular segment whose radius is R′, and abscissa x, Art. (88). Integrating this between the limits $x = 0$, and $x = \mathrm{CB} = a$, and calling the segment CBFG = D, we have

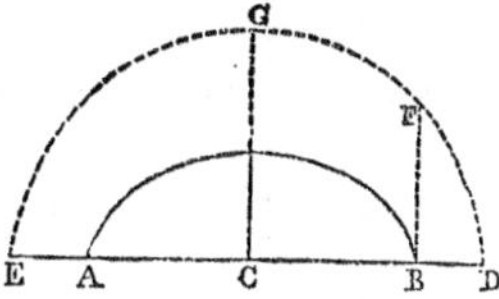

$$u'' = \mathrm{C}'\mathrm{D} = \frac{1}{2}\text{ area of Ellipsoid.}$$

If $a = b$ in the primitive value of u, we shall have

$$u = \int 2\pi a dx = 2\pi a x + C,$$

for the surface of the circumscribing sphere.

Let the area of a paraboloid of revolution be determined.

214′ By the substitution of the value of dx, Art. (201), in the general expression for u, we have for the surface generated by the revolution of a cycloid about its base,

$$u = 2\pi \sqrt{2r} \int y dy (2r - y)^{-\frac{1}{2}}.$$

Placing $2r - y = z$, and integrating as in Art. (131), we have

$$u = 2\pi\sqrt{2r}\left(-4r(2r-y)^{\frac{1}{2}} + \frac{2}{3}(2r-y)^{\frac{3}{2}}\right) + C.$$

Taking the area between the limits $y = 0$, and $y = 2r$, we have

$$u'' = \frac{32}{3}\pi r^2,$$

for one-half the surface. The whole is $\frac{64}{3}$ *the area of the generating circle.*

CUBATURE OF SOLIDS OF REVOLUTION.

215. The operation by which the solid content, or solidity of a solid, is determined, is called *its cubature.*

For the differential of a solid of revolution we have found, Art. (90),

$$dv = \pi y^2 dx, \qquad \text{or} \qquad v = \int \pi y^2 dx \ldots\ldots\ldots\ldots(1);$$

in which y and x represent the co-ordinates of the curve which generates the bounding surface, the axis of X being the axis of revolution.

For the cubature of any particular solid; *we find, from the equation of its meridian curve, the value of* y^2 *in terms of* x; *or from the differential equation of the curve, the value of* dx *in terms of* y *and* dy, *and substitute in the above formula* (1); *the result of the integration will be an expression for an indefinite portion of the solid.*

216. Let the rectangle ABCD revolve about AB and generate a right cylinder. The origin of co-ordinates being at A, the equation of DC will be,

$$y = \text{AD} = b,$$

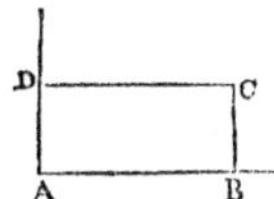

then

$$v = \int \pi y^2 dx = \int \pi b^2 dx = \pi b^2 x + \text{C}.$$

Taking this between the limits $x = 0$, and $x = \text{AB} = h$, we have

$$v'' = \pi b^2 h = \textit{the base into the altitude.}$$

217. The equation of the ellipse gives

$$y^2 = \frac{b^2}{a^2}(a^2 - x^2);$$

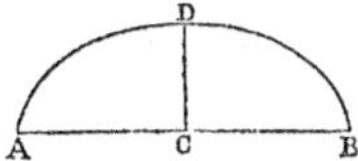

whence for the ellipsoid of revolution

$$v = \int \pi \frac{b^2}{a^2}(a^2 - x^2)dx = \frac{\pi b^2}{a^2}(a^2 x - \frac{x^3}{3}) + \text{C}.$$

Estimating the solidity from the plane through the centre, perpendicular to the transverse axis, we have $x = 0$, $C = 0$, and

$$v' = \frac{\pi b^2}{a^2}(a^2 x - \frac{x^3}{3}).$$

Making $x = a$, we obtain for one half the solid

$$v'' = \frac{\pi b^2}{a^2}(a^3 - \frac{a^3}{3}) = \frac{2}{3}\pi b^2 a,$$

and for the whole

$$\frac{4}{3}\pi b^2 a = \frac{2}{3}\pi b^2 \times 2a;$$

or, *equal to two-thirds of the circumscribing cylinder.*

If the same ellipse revolve about its conjugate axis, we have

$$v = \int \pi x^2 dy = \int \pi \frac{a^2}{b^2}(b^2 - y^2)dy,$$

which between the limits $y = -b$ and $y = b$, gives

$$v'' = \frac{4}{3}\pi a^2 b = \frac{2}{3}\pi a^2 \times 2b.$$

The latter solid is called the *oblate spheroid*, and the former the *prolate spheroid*, and we have the proportion

$$\text{the prolate : the oblate} :: \frac{4}{3}\pi b^2 a : \frac{4}{3}\pi a^2 b :: b : a.$$

If in either expression $a = b$, we have

$$\frac{4}{3}\pi a^3 = \textit{solidity of a sphere.}$$

Let the origin be now taken at A, when

$$y^2 = \frac{b^2}{a^2}(2ax - x^2),$$

and the solidity be determined.

Give also the cubature of a sphere directly, by using the equation

$$y^2 + x^2 = r^2.$$

218. Give also the cubatures of the following solids of revolution:

1. The right cone, $v'' =$ base $\times$ $\frac{1}{3}$ of altitude.

2. The paraboloid, $v'' = \frac{1}{2}$ circumscribing cylinder.

3. The solid generated by a given portion of the common parabola revolving about the tangent at its vertex,
$v'' = \frac{1}{5}$ cylinder with same base and altitude.

4. The solid, the bounding surface of which is generated by the curve whose equation is $y^2 = \frac{a}{x}$.

5. The solid, the bounding surface of which is generated by one branch of the cycloid revolving about its base.

APPLICATION OF THE CALCULUS TO SURFACES.

219. Since the equation of every surface expresses the relation between the co-ordinates of its points, it must contain three variables, and may be generally written

$$u = \mathrm{F}(x, y, z) = 0 \ldots\ldots(1);$$

or since either two of these variables may be assumed at pleasure, and the remaining one determined from the equation, the latter may be regarded as a function of the other two, they being entirely independent of each other, and the equation of the surface be thus otherwise expressed,

$$z = f(x, y) \ldots\ldots(2).$$

In the equation of every surface considered, z will be regarded as a function of x and y; and the co-ordinate planes will be taken at right angles to each other.

The differential equation of a surface may then be obtained, either by differentiating equation (1), as in article (54), or by differentiating equation (2), as in article (49). By the latter method we obtain

$$dz = \frac{dz}{dx}dx + \frac{dz}{dy}dy \ldots\ldots\ldots\ldots(3).$$

220. Let M be any point of a surface, a portion of which is represented in the annexed figure. The co-ordinates of this point are

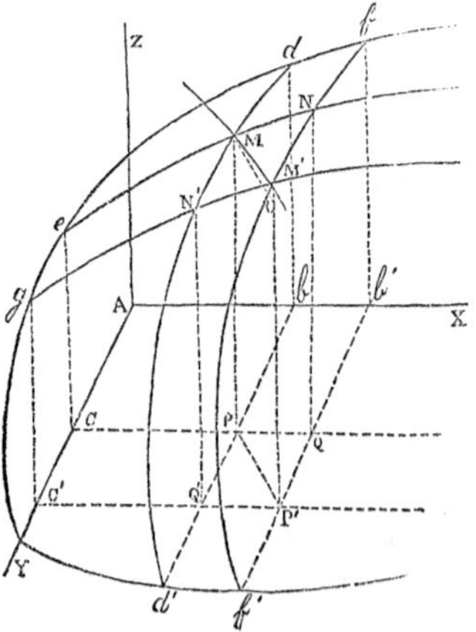

$$x = \mathrm{A}b, \quad y = \mathrm{AC}, \quad z = \mathrm{MP}.$$

Let a plane be passed through M parallel to YZ. For every point of this plane

$$x = \mathrm{A}b = x''.$$

If, then, in the equation of the surface, we make $x = x''$, and suppose z and y to vary, they can only belong to points in the curve dMd', the intersection of the plane and surface; and if we suppose y to receive the increment $CC' = k$, we shall have, by Taylor's formula,

$$N'Q' = f(x'', y + k) = z + \frac{dz}{dy}k + \frac{d^2z}{dy^2}\frac{k^2}{1.2} + \&c.,$$

in which x is regarded as constant and equal to x''.

In the same way, if $y = y''$ in the equation of the surface. and z and x vary, we shall have the curve eMN, and if x receive the increment $bb' = h$,

$$NQ = f(x + h, y'') = z + \frac{dz}{dx}h + \frac{d^2z}{dx^2}\frac{h^2}{1.2} + \&c. \ldots.$$

If now x and y at the same time receive the increments h and k respectively, we have, Art. (46),

$$M'P' = z' = f(x + h, y + k) = z + \frac{dz}{dx}h + \frac{d^2z}{dx^2}\frac{h^2}{1.2} + \&c.$$

$$+ \frac{dz}{dy}k + \frac{d^2z}{dxdy}hk + \&c.$$

$$+ \frac{d^2z}{dy^2}\frac{k^2}{1.2} + \&c.$$

or

$$z' - z = ph + p'k + \frac{1}{2}(qh^2 + 2q'hk + q''k^2) + \&c.;$$

by making

$$\frac{dz}{dx} = p, \qquad \frac{dz}{dy} = p', \qquad \frac{d^2z}{dx^2} = q\ldots\ldots\&c.$$

When $x = x''$, equation (3) gives

$$dz = \frac{dz}{dy}dy = p'dy, \qquad \text{or} \qquad \frac{dz}{dy} = p' \ldots\ldots (4),$$

equations which evidently belong only to the section dMd' parallel to YZ.

If $y = y''$, the corresponding equations for the section parallel to XZ are

$$dz = \frac{dz}{dx}dx = pdx, \qquad \text{or} \qquad \frac{dz}{dx} = p \ldots\ldots (5).$$

The value of $\frac{dz}{dy}$, equation (4), is the tangent of the angle which a tangent to the section dMd', at any point, makes with the axis of Y, or with the plane XY; and $\frac{dz}{dx}$, equation (5), the corresponding expression for the section eMN; and since these angles are the same as those made by the curves at the point of contact, with XY, they give the inclination or *slope* of the surface in the direction of these curves.

221. If it be required to find the slope of the surface at any point, as M, along the section MM′ made by the plane MM′PP′, we take the equation of this plane

$$y = ax + \beta \ldots\ldots (1), \qquad z \textit{ indeterminate};$$

a being the tangent of the angle made with the axis of X by the trace PP′, and equal to $\frac{dy}{dx} = \frac{k}{h}$.

Now in order that z shall represent only the ordinates of points in the section MM′, the relation expressed in equation (1) must exist between the variables x and y, and we must have

$$dy = \alpha dx,$$

which in equation (3) of article (219), gives

$$dz = (p + \alpha p')dx.$$

The limit of the ratio $\dfrac{\text{M}'\text{P}' - \text{MP}}{\text{PP}'}$ is evidently the tangent of the angle (S) which the tangent, and consequently the curve at the point M, makes with PP′, or with the plane XY.

But since

$$\text{PP}' = \sqrt{\overline{\text{P}'\text{Q}}^2 + \overline{\text{PQ}}^2} = h\sqrt{1 + \alpha^2},$$

we have

$$\frac{\text{M}'\text{P}' - \text{MP}}{\text{PP}'} = \frac{z' - z}{h\sqrt{1 + \alpha^2}},$$

the limit of which is

$$\frac{1}{\sqrt{1 + \alpha^2}} \times \frac{dz}{dx} = \frac{p + \alpha p'}{\sqrt{1 + \alpha^2}} = \text{tang S}.$$

To find the direction in which the section MM′ must be made in order that the slope at a given point M, along the curve cut out, be greater than along any other, it is only necessary to obtain that value of α which will render the expression

$$\frac{p + \alpha p'}{\sqrt{1 + \alpha^2}},$$

a maximum, the values of p and p' being taken at the given point

M. Differentiating the expression with reference to α, and placing the result equal to 0, we have

$$\frac{p' - p\alpha}{(1 + \alpha^2)^{\frac{3}{2}}} = 0;$$

whence

$$p' - p\alpha = 0, \qquad \alpha = \frac{p'}{p}.$$

This value of α substituted in equation (1), (β being first determined by the condition that the line PP′ shall pass through P), will give an equation, which, combined with that of the surface, will determine the line of *greatest slope.*

222. The co-ordinates of a given point M, being x'', y'', and z''; the equations of a tangent to the section parallel to XZ at this point, will be

$$z - z'' = m(x - x''), \qquad y = y'';$$

and to the section parallel to YZ,

$$z - z'' = n(y - y''), \qquad x = x'';$$

in which m and n represent what $\frac{dz}{dx}$ and $\frac{dz}{dy}$, equations (5) and (4) of article (220), become, when x'', y'' and z'' are substituted for x, y and z.

The line, of which the equations are

$$z - z'' = -\frac{1}{m}(x - x''), \qquad z - z'' = -\frac{1}{n}(y - y''),$$

is perpendicular to both of these tangents, and, of course, to their plane, which is tangent to the surface. This line is then a normal to the surface at the given point. The equation of a plane passing through this point is

$$A(x - x'') + B(y - y'') + C(z - z'') = 0.$$

To make this plane tangent to the surface, it is necessary to introduce into its equation the conditions that it be perpendicular to the normal, which are

$$A = -mC, \qquad B = -nC;$$

whence

$$-m(x - x'') - n(y - y'') + (z - z'') = 0,$$

or substituting for m and n their values, $\frac{dz''}{dx''}$ and $\frac{dz''}{dy''}$, and reducing

$$\frac{dz''}{dx''}(x - x'') + \frac{dz''}{dy''}(y - y'') - (z - z'') = 0 \ldots\ldots(1).$$

The equation and differential equation of the Ellipsoid are

$$Mz^2 + Ny^2 + Lx^2 - P = 0, \quad \text{and} \quad Mzdz + Nydy + Lxdx = 0;$$

whence

$$\frac{dz''}{dx''} = -\frac{Lx''}{Mz''}, \qquad \frac{dz''}{dy''} = -\frac{Ny''}{Mz''},$$

which in equation (1), after reduction, give

$$Lx''(x - x'') + Ny''(y - y'') + Mz''(z - z'') = 0,$$

or since

$$- Lx''^2 - Ny''^2 - Mz''^2 = - P,$$

$$Mzz'' + Nyy'' + Lxx'' - P = 0,$$

for the tangent plane to the ellipsoid at a given point.

If $M = N = L$, we have, for the tangent plane to the sphere,

$$zz'' + yy'' + xx'' - R^2 = 0.$$

The distance from any point of the normal to the point of contact is,

$$D = \sqrt{(x - x'')^2 + (y - y'')^2 + (z - z'')^2}$$

$$= (z'' - z)\sqrt{1 + m^2 + n^2}.$$

If $z = 0$, we have

$$D = z''\sqrt{1 + m^2 + n^2},$$

for the distance from the point of the normal in the plane XY, to the point of contact.

223. One surface is osculatory to another, when it has with it a more intimate contact than any other surface of the same kind; and the conditions which must exist in order that a surface, given in kind only, shall be osculatory to a given surface at a given point, can be determined by a method similar to that pursued in article (97). But from the nature of the case these conditions are more numerous and complicated, and their determination more difficult; so much so as to render osculatory surfaces of little use in the measure of curvature; hence another method has been devised which will now be explained.

Let M be any point of a surface, at which it is proposed to examine the curvature. Let this point be taken as the origin of co-ordinates, and let the normal at this point be the axis of Z, the axes of X and Y having any position in the tangent plane XMY. The equation of the surface, Art. (219), will be

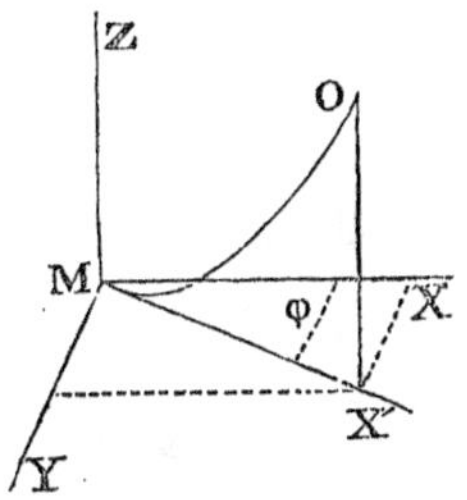

$$z = f(x, y)\text{............}(1).$$

Through the normal let any plane ZMX′, making an angle φ with the plane ZX, be passed; it will cut from the surface a curve MO. For any point of this curve, as O, denoting the abscissa MX′ by x', we shall have

$$x = x' \cos\varphi, \qquad y = x' \sin\varphi\text{............}(2),$$

and these values, substituted in equation (1), will evidently give the equation of the curve referred to the two axes MZ and MX′. Now by varying the angle φ, all the normal sections at the point M may be obtained, and by examining the curvatures of these different sections at the given point, an accurate idea of the curvature of the surface may be formed.

The general expression for the radius of curvature of one of these sections, Art. (105), may be put under the form

$$R = \frac{\left(1 + \frac{dz^2}{dx'^2}\right)^{\frac{3}{2}}}{\frac{d^2z}{dx'^2}}\text{............}(3).$$

Differentiating equations (2), we have

$$dx = dx' \cos\varphi, \qquad dy = dx' \sin\varphi\text{............}(4);$$

and substituting these values in equation (3), Art. (219),

$$dz = p \cos \varphi dx' + p' \sin \varphi dx', \text{ or } \frac{dz}{dx'} = p \cos \varphi + p' \sin \varphi \ldots\ldots(5).$$

Differentiating again, recollecting that p and p' are implicit functions of x', we have, Art. (220),

$$\frac{d^2z}{dx'^2} = \cos \varphi \left(q\frac{dx}{dx'} + q'\frac{dy}{dx'} \right) + \sin \varphi \left(q'\frac{dx}{dx'} + q''\frac{dy}{dx'} \right);$$

or since equations (4) give $\frac{dx}{dx'} = \cos \varphi$ and $\frac{dy}{dx'} = \sin \varphi$,

$$\frac{d^2z}{dx'^2} = q \cos^2\varphi + 2q' \cos\varphi \sin\varphi + q'' \sin^2\varphi \ldots\ldots(6).$$

If these values of $\frac{dz}{dx'}$ and $\frac{d^2z}{dx'^2}$ be substituted in expression (3), we shall have the general value for the radius of curvature of any one of the normal sections. But as we only desire this value for the point M, we may first substitute the co-ordinates of this point, which are

$$x'' = 0, \qquad y'' = 0, \qquad z'' = 0;$$

and since the normal at this point coincides with the axis of Z, we must also have, Art. (222)

$$\frac{dz''}{dx''} = 0, \qquad \frac{dz''}{dy''} = 0, \qquad \text{or} \qquad p = 0, \quad p' = 0.$$

substituting these values in equations (5) and (6), and the results in equation (3), we obtain

$$R = \frac{1}{q \cos^2\varphi + 2q' \cos\varphi \sin\varphi + q'' \sin^2\varphi} \ldots\ldots\ldots\ldots(7),$$

in which q, q' and q'' are what the partial differential coefficients of the second order of the function z become, when 0 is substituted for x, y and z.

Dividing by $\cos^2 \varphi$ and recollecting that $\dfrac{1}{\cos^2 \varphi} = 1 + \text{tang}^2 \varphi$, this value may be put under the form

$$\text{R} = \frac{1 + \text{tang}^2 \varphi}{q + 2q' \text{tang}\, \varphi + q'' \text{tang}^2 \varphi} \ldots\ldots\ldots\ldots (8).$$

We have taken the positive value of R, Art. (105), since, as the surface is represented in the figure, the sections are above the axis of X′ and convex towards it; $\dfrac{d^2 z}{dx'^2}$ must therefore be positive, Art. (83), and the value of R positive, as it should be when laid off from M above the plane XY. If the section at the point M lies below the plane XY, it must still be convex towards this tangent plane; $\dfrac{d^2 z}{dx'^2}$ will be negative, and R negative, and must therefore be laid off from M below XY.

By assigning all values to φ from 0 to 360° in equation (8), we shall obtain a value of R for each normal section. Among these values there must be one which is greater, and another which is less than all the others. The values of φ which will give these principal values of R will be obtained as in Art. (66).

Differentiating equation (8), we have

$$\frac{d\text{R}}{d \,\text{tang}\, \varphi} = \frac{2(q' \text{tang}^2 \varphi + (q - q'') \text{tang}\, \varphi - q')}{(q + 2q' \text{tang}\, \varphi + q'' \text{tang}^2 \varphi)^2}.$$

If the denominator be placed equal to 0, we shall obtain values of the tang φ, which, when real, will reduce the value of R to infinity. The curvature of the corresponding section will then be zero, and the section itself a right line, or the point M a singular point, Art. (92), cases which do not occur in all surfaces. Let us then place the numerator equal to 0, we thus have

$$\text{tang}^2 \varphi + \frac{q - q''}{q'} \text{tang}\, \varphi - 1 = 0 \ldots\ldots\ldots\ldots (9).$$

This being either of the first or second form of equations of the second degree, the roots will always be real and their product equal to -1, that is, denoting them by tang φ' and tang φ''

$$\text{tang}\, \varphi' \, \text{tang}\, \varphi'' + 1 = 0\,;$$

hence the normal planes in which the greatest and least radii of curvature are found, must be perpendicular to each other. Their exact position will be determined by solving equation (9).

The values of tang φ' and tang φ'' being determined and the traces of the normal planes constructed as in the figure; let us take MX$''$ as a new axis of X, and MY$''$ as a new axis of Y, and suppose the surface to be referred to them with MZ as an axis of Z. Then we must have for these new axes

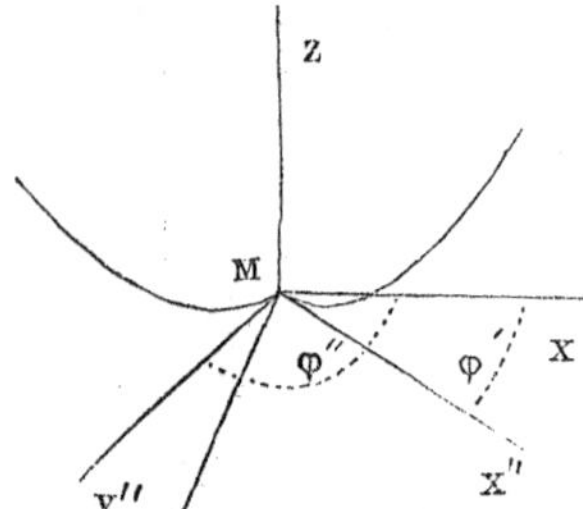

$$\text{tang}\, \varphi' = 0, \quad \text{tang}\, \varphi'' = \infty, \quad \text{tang}\, \varphi' + \text{tang}\, \varphi'' = \infty,$$

which requires in equation (9) that $q' = 0$. Substituting this value of q' in equation (7), we have

$$R = \frac{1}{q \cos^2 \varphi + q'' \sin^2 \varphi} \ldots\ldots\ldots\ldots (10).$$

Substituting in this the values of φ, corresponding to the maximum and minimum radii as above determined, viz. $\varphi = 0$ and $\varphi = 90°$, and denoting the values of the principal radii thus determined by R$'$ and R$''$, we have

$$R' = \frac{1}{q}, \qquad R'' = \frac{1}{q''},$$

and finally from equation (10),

$$\frac{1}{R} = q \cos^2 \varphi + q'' \sin^2 \varphi = \frac{1}{R'} \cos^2 \varphi + \frac{1}{R''} \sin^2 \varphi,$$

which expresses the reciprocal of the radius of curvature of any normal section, in terms of the principal radii and the angle φ.

If R' and R'' are both positive, all values of R will be positive, and the greatest of the two will be a maximum, and the least a minimum, and all the normal sections at the point M will lie above the plane XY.

If R' and R'' are both negative, the sections will lie below XY. If one is positive and the other negative, a part of the values of R will be positive and a part negative, and a part of the sections will be above and a part below the plane XY, and this plane will cut the surface at the point M, giving a point analogous to the point of inflexion, Art. (92).

If $R' = R''$, all the values of R become equal to R' or R'', and the curvature of all the sections will be the same; as at any point of a sphere, or at the vertex of a surface of revolution.

224. To determine a general expression for the solidity of any solid; denote the solid AbPc-MZ, included by the surface, the co-ordinate planes and the parallels ece' and dbd', by v. Since, by the equation of the bounding surface, z will always be given in terms of x and y, the solid may be regarded as a function of x and y. Let x be increased by h, y remaining the same, we shall have the solid

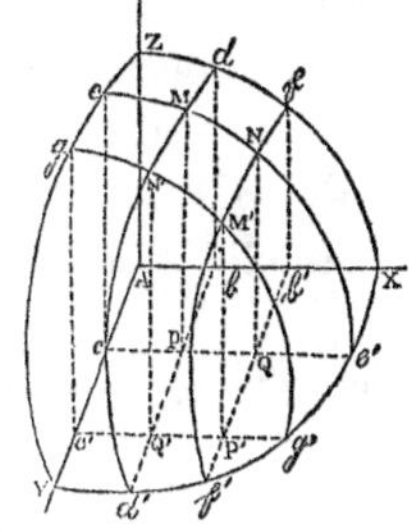

$$bb'\text{QP-N}d = v' - v = \frac{dv}{dx}h + \frac{d^2v}{dx^2}\frac{h^2}{1.2} + \&c.$$

If y be increased by k, and x remain unchanged, we shall have the solid

$$c\text{PQ}'c'\text{-N}'e = v'' - v = \frac{dv}{dy}k + \frac{d^2v}{dy^2}\frac{k^2}{1.2} + \&c.$$

If now x and y be increased at the same time by the variables h and k respectively, we shall have the solid whose base is $c\text{P}bb'\text{P}'c'$, or

$$v''' - v = +\frac{dv}{dx}h + \frac{d^2v}{dx^2}\frac{h^2}{1.2} + \&c.$$

$$+\frac{dv}{dy}k + \frac{d^2v}{dxdy}hk + \&c.$$

$$+\frac{d^2v}{dy^2}\frac{k^2}{1.2} + \&c.$$

Subtracting the sum of the first two, from the last, we have

$$\textit{solid}\ \text{PQP}'\text{Q}' - \text{MM}' = \frac{d^2v}{dxdy}hk + \frac{d^3v}{1.2dx^2dy}h^2k + \&c.$$

If through M and M′, planes be passed parallel to XY, two parallelopipedons will be formed, having the common base PQP′Q′ and the altitudes MP and M′P′; the limit of the ratio of these solids will evidently be equal to unity, and since the solid PQP′Q′ - MN is always less than one and greater than the other, the limit of its ratio to either will also be unity, Art. (85).

The solidity of the first parallelopipedon being hk.MP, we have the ratio

$$\frac{\frac{d^2v}{dxdy}hk + \frac{d^3v}{dx^2dy}\frac{h^2k}{1.2} + \&c.}{hk.\text{MP}} = \frac{\frac{d^2v}{dxdy} + \frac{d^3v}{dx^2dy}\frac{h}{1.2} + \&c.}{z};$$

and passing to the limit

$$\text{L} = \frac{\frac{d^2v}{dxdy}}{z} = 1; \qquad \frac{d^2v}{dxdy} = z;$$

or

$$\frac{d\left(\frac{dv}{dy}\right)}{dx} = z \qquad d\left(\frac{dv}{dy}\right) = zdx.$$

Integrating with respect to x,

$$\frac{dv}{dy} = \int zdx + \text{Y}.$$

From this

$$dv = dy\int zdx + \text{Y}dy.$$

Integrating both members with reference to y,

$$v = \int dy \int zdx + \int \text{Y}dy + \text{X},$$

or Art. (164),

$$v = \int^2 zdydx + \int \text{Y}dy + \text{X}.$$

Since the integral $\int zdx + \text{Y}$ is evidently the area of one of the parallel sections as eMe'; to obtain the whole solidity represented in the figure, we must first take the integral between the limits

$x = 0$ and $x = ce'$, and then the second integral between the limits $y = 0$ and $y = \text{AY}$.

To illustrate, let us determine the solidity of the pyramid ABD - C; the equation of the plane BDC, being

$$x + 2y + 3z - 2 = 0;$$

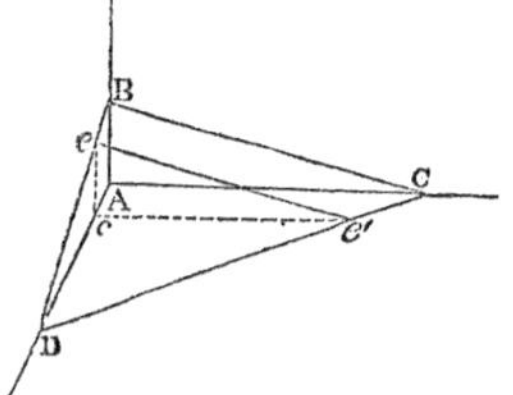

whence

$$z = \frac{2 - 2y - x}{3}.$$

The equation of DC is

$$x + 2y = 2, \qquad \text{or} \qquad x = 2 - 2y,$$

$$\text{AD} = 1, \qquad \text{AC} = 2, \qquad \text{AB} = \frac{2}{3},$$

$$v = \int^2 z dx dy = \int dy \int dx \frac{(2 - 2y - x)}{3}.$$

Integrating with respect to x,

$$v = \int dy \left(\frac{2x - 2yx}{3} - \frac{x^2}{6} + \text{Y} \right),$$

or taking the integral between the limits

$$x = 0 \qquad \text{and} \qquad x = ce' = 2 - 2y,$$

$$v = \int dy \frac{(4 - 8y + 4y^2)}{6}.$$

Integrating now with reference to y, between the limits

$$y = 0 \qquad \text{and} \qquad y = \text{AD} = 1,$$

we obtain for the solidity

$$v = \frac{4}{18} = \frac{1}{2} \times \frac{2}{3} \times 1 \times \frac{2}{3} = \frac{1}{2}\,\text{AB} \times \text{AD} \times \frac{1}{3}\,\text{AC}$$

$$= \text{BAD} \times \frac{1}{3}\,\text{AC}.$$

225. Let BMM′ be any curve in space, and B′PP′ its projection on the co-ordinate plane XY. Let the plane of the curve MM′ make an angle β with the plane XY, and let its intersection with that plane be taken for the axis of X. Then, if the ordinate MQ be denoted by y', the area of the curve MM′ will be

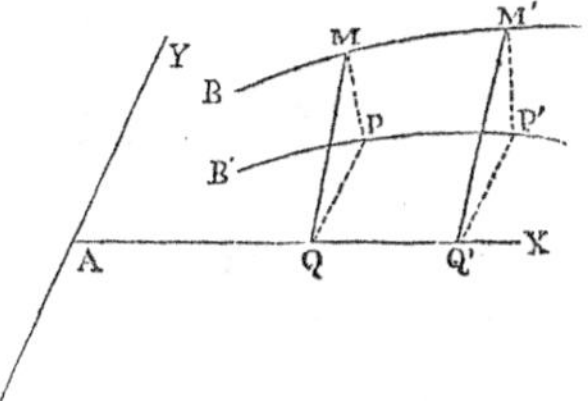

$$s = \int y'dx \ldots\ldots\ldots \text{Art. (203)}.$$

But any ordinate PQ of the projection is plainly equal to the corresponding ordinate MQ of the curve multiplied by $\cos \text{MQP} = \cos \beta$, or

$$y = y' \cos \beta\,;$$

hence the area of the projection B′PP′, denoted by S, is

$$\text{S} = \int y dx = \int y' \cos \beta \, dx = \cos \beta \int y'dx = \cos \beta \, s\,;$$

that is, *the projection of any plane area is equal to the area multiplied by the cosine of the angle included between its plane and the plane of projection.*

226. Now, let u denote the area of any curved surface. It will be a function of x and y. By a process identical with that of Art. (224), we shall find the surface

$$\text{MNM'N'} = \frac{d^2u}{dxdy}hk + \frac{d^3u}{dx^2dy}\frac{h^2k}{1.2} + \&c. \ldots\ldots(1).$$

If a tangent plane be drawn at M, and the four planes PN, QM′, &c. be produced, they will form on the tangent plane the parallelogram MORS. The limit of the ratio of this parallelogram and the surface MNM′N′ will be unity, as may be proved by a process similar to that pursued in article (89).

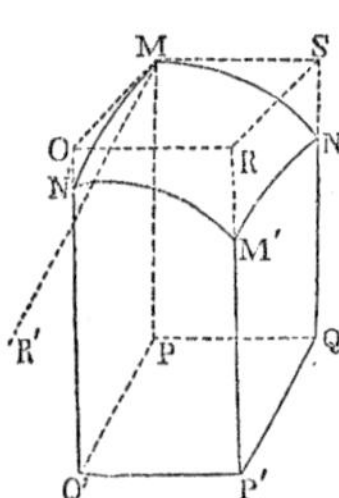

The area of the parallelogram is equal to its projection PQP′Q′ divided by $\cos\beta$; β being the angle which the tangent plane makes with XY. But β is also the angle which the normal MR′ makes with MP or the axis of Z; hence

$$\cos\beta = \frac{1}{\sqrt{1+m^2+n^2}},$$

$-m$ and $-n$ representing the tangents $\left(-\frac{dz}{dx} \text{ and } -\frac{dz}{dy}\right)$ of the angles which the projections of MR′ make with the axis of Z, Art. (222); hence

$$area\ \text{MORS} = \frac{\text{PQP'Q'}}{\cos\beta} = \frac{hk}{\frac{1}{\sqrt{1+m^2+n^2}}} = hk\sqrt{1+m^2+n^2}.$$

Dividing equation (1) by this, we have

$$\frac{\text{MNM'N'}}{\text{MORS}} = \frac{\dfrac{d^2u}{dxdy} + \dfrac{d^3u}{dx^2dy}\dfrac{h}{1.2} + \&c.}{\sqrt{1 + m^2 + n^2}}$$

Passing to the limit, we have

$$\text{L} = \frac{\dfrac{d^2u}{dxdy}}{\sqrt{1 + m^2 + n^2}} = 1;$$

whence

$$d^2u = dxdy\sqrt{1 + m^2 + n^2},$$

and

$$u = \int^2 dxdy\sqrt{1 + m^2 + n^2}.$$

For the sphere, we have

$$x^2 + y^2 + z^2 = \text{R}^2;$$

whence

$$\frac{dz}{dx} = -\frac{x}{z} = \frac{-x}{\sqrt{\text{R}^2 - x^2 - y^2}} = m,$$

$$\frac{dz}{dy} = -\frac{y}{z} = \frac{-y}{\sqrt{\text{R}^2 - x^2 - y^2}} = n,$$

$$\sqrt{1 + m^2 + n^2} = \frac{\text{R}}{\sqrt{\text{R}^2 - x^2 - y^2}},$$

and

$$u = \int^2 \frac{\text{R}dxdy}{\sqrt{\text{R}^2 - x^2 - y^2}}.$$

Making $\sqrt{R^2 - y^2} = R'$, and integrating with reference to x, we have

$$u = \int R dy \int \frac{dx}{\sqrt{R'^2 - x^2}} = \int R dy \sin^{-1}\frac{x}{R'}$$

$$= \int R dy \left(\sin^{-1}\frac{x}{\sqrt{R^2 - y^2}} + Y\right).$$

Taking the integral between the limits,

$$x = 0 \quad \text{and} \quad x = ce' = \sqrt{R^2 - y^2},$$

we have

$$u = \int R dy \frac{\pi}{2}.$$

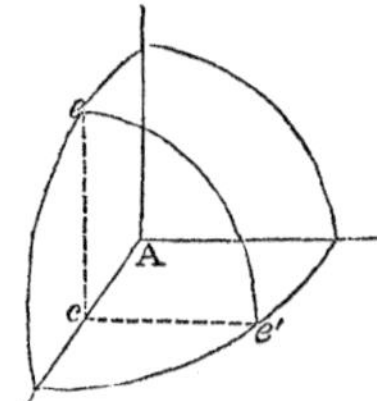

Integrating again, with reference to y, we have

$$u = \frac{R\pi}{2}y + C,$$

and between the limits $y = 0$, $y = R$,

$$u'' = \frac{\pi R^2}{2},$$

for one-eighth of the surface. The entire surface is then

$$4\pi R^2.$$

PART III.

CALCULUS OF VARIATIONS.

FIRST PRINCIPLES.

227. A function may be regarded as given, when the form of the algebraic expression, which determines the relation between it and the variable or variables, is given, and the constants which enter this expression are known.

In this case, the only change which the function can be made to undergo, is that which arises from a change in the variables. When these variables receive infinitely small increments, the corresponding infinitely small increment or change of the function is taken for *the differential of the function*, Art. (91). All our previous applications of the Calculus have been made to functions of the kind above referred to, and the term differential can, with propriety, be applied to no other change.

It will at once be seen, that if a function be not given as above described, but *merely subjected to certain conditions*, it may be made to undergo a change by altering the relation which exists between it and the variables; and this may be done by changing either the form of the expression or the constants which enter it, in any way consistent with the given conditions. Now if such a change be made as to give *another function consecutive with the first*, the infinitely small change which the first undergoes is called *its variation*, and the corresponding changes of the variables are their *variations*.

The difference between the terms "differential" and "variation," will be made more plain by geometrical illustration.

Let BC be any curve, a function of x and y, of which M and M′ are any two consecutive points, the co-ordinates of M being x and y. Now if the constants which determine the curve be changed in any way so as to give a different curve B′C′, *infinitely near* to BC, and so that the points M and M′ shall take the positions m and m', Pp will be the variation of x and mS the variation of y, while PP′ is the differential of x and M′Q the differential of y, Art. (91).

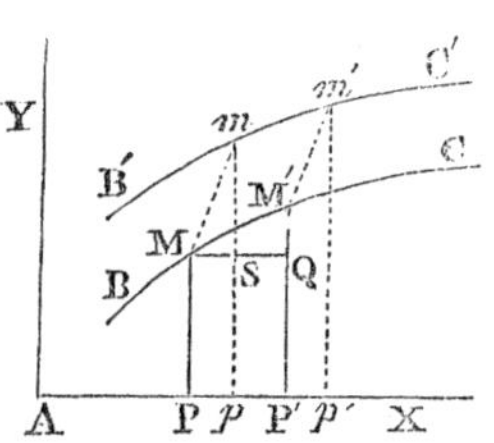

The conditions under which the variation is made, may be such that one of the variables will have no variation; and when this is the case, the operations to be performed will be much simplified: Thus, if it be required that the points M and M′ shall be found in lines parallel to the axis of Y at m and m', Mm will be the variation of y, while x has no variation; the differentials of x and y being PP′ and M′Q as before.

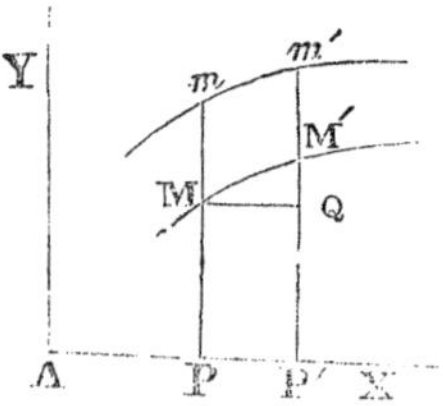

As the differential is denoted by the symbol d, the Greek character δ is used to denote the variation, and from the illustrations just given, it appears that while the former symbol denotes the changes which take place in passing from *one point to another of the same curve* the latter is used for a very different purpose, to denote the changes in passing from points of one curve to the corresponding points of another *infinitely near to it.*

228. From the nature of the term as above explained, we see that to obtain the variation of any function of x, y, z, &c., we have only to put for x, y, z, &c., $x + \delta x$, $y + \delta y$, &c., and then take, as in the Differential Calculus, Art. (49), those terms of the development which are of the first degree with reference to the variations of the variables: Or, since the development may be made precisely as in Art. (51), by substituting δx, δy, &c., for h, k, &c., it is plain that we shall have

$$\delta u = \frac{du}{dx}\delta x + \frac{du}{dy}\delta y + \frac{du}{dz}\delta z + \text{\&c.}$$

It is also plain that the principles contained in articles (13) and (15), as also the particular rules demonstrated in articles (18) ... (24), are equally applicable to variations.

229. In the function

$$u = f(x) \ldots\ldots\ldots\ldots (1),$$

let us substitute $x + \delta x$ for x, and denote the new function by $f'(x)$; then by the definition, Art. (227),

$$\delta u = f'(x) - f(x) \ldots\ldots\ldots\ldots (2),$$

and since from the relation expressed in equation (1), x is a function of u, the second member of equation (2) will be a function of u, and we may write

$$\delta u = \varphi(u) \ldots\ldots\ldots\ldots (3).$$

If in this equation we put for u, $u + du = u'$, we shall have

$$\delta u' = \varphi(u');$$

subtracting equation (3)

$$\delta u' - \delta u = \varphi(u') - \varphi(u) = d\varphi(u) = d\delta u.$$

But

$$u' - u = du,$$

and taking the variations, Arts. (13) and (18), we have

$$\delta u' - \delta u = \delta du;$$

hence

$$\delta du = d\delta u \ldots\ldots\ldots\ldots (4).$$

That is; *the variation of the differential of a function, is equal to the differential of its variation:* Or when both of the symbols d and δ are prefixed to a function, the order in which they are written, or in which the operations indicated are performed, can be changed at pleasure without affecting the result.

The principle above enunciated is true for any order of the differential; for if in equation (4) we put du for u, we have

$$\delta d(du) = d\delta du \qquad \text{or} \qquad \delta d^2u = dd\delta u = d^2\delta u.$$

If in the last equation we put du for u, we have

$$\delta d^2(du) = d^2\delta du, \qquad \text{or} \qquad \delta d^3u = d^3\delta u,$$

and so on; hence we may conclude that,

$$\delta d^n u = d^n\delta u.$$

230. Let v be any differential of a function of x, and place

$$\int v = v', \qquad \text{then} \qquad dv' = v,$$

$$\delta dv' = \delta v, \qquad \text{or} \qquad d\delta v' = \delta v,$$

and by integration

$$\delta v' = \int \delta v, \qquad \text{or} \qquad \delta \int v = \int \delta v.$$

The principles demonstrated in this and the preceding article, are evidently true for functions of any number of variables; since the variation of the differential of such a function is but the sum of the partial variations, and the converse.

231. In order to consider the subject of variations in its most general sense, when applied to differential expressions, we must regard the differentials of all the variables as variable, as well as the variables themselves. In this sense, if u be a function containing x, y, and their successive differentials, we shall have, Art. (228),

$$\left.\begin{aligned}\delta u = \mathrm{M}\delta x + \mathrm{M}'\delta dx + \mathrm{M}''\delta d^2x + \&c. \\ + \mathrm{N}\delta y + \mathrm{N}'\delta dy + \mathrm{N}''\delta d^2y + \&c.\end{aligned}\right\} \ldots\ldots\ldots(1).$$

in which M, M,′ M″ &c. are the partial differential coefficients of u taken with respect to x, dx, d^2x, &c.; and N, N′, N″ &c., the corresponding ones taken with respect to y, dy, d^2y, &c.; and if this expression be first extended to any number of variables, by adding for each an expression of the form

$$\mathrm{M}\delta x + \mathrm{M}'\delta dx + \mathrm{M}''\delta d^2x + \&c.$$

it may then be made to give every particular case which can arise, by making the particular suppositions upon dx, d^2x, dy, d^2y, &c., which the case requires.

We often meet with differential expressions containing only the variables x, y, $\frac{dy}{dx} = p$, $\frac{d^2y}{dx^2} = q$, &c. If we denote such expression by v, we shall have, as in Art. (228),

$$\delta v = \mathrm{M}\delta x + \mathrm{N}\delta y + \mathrm{N}'\delta p + \mathrm{N}''\delta q + \&c. \ldots\ldots (2).$$

And if this expression be taken in its most general sense, dx must be regarded as variable, in which case we put for δp, δq, &c., their values obtained as in Art. (24), viz.,

$$\delta p = \delta\frac{dy}{dx} = \frac{dx\delta dy - dy\delta dx}{dx^2} = \frac{d\delta y - pd\delta x}{dx},$$

$$\delta q = \delta\frac{dp}{dx} = \frac{dx\delta dp - dp\delta dx}{dx^2} = \frac{d\delta p - qd\delta x}{dx}.$$

If dx be regarded as constant, equation (2) is under its most simple form.

232. If u be still regarded in its most general sense, we have, Art. (230),

$$\delta\int u = \int\delta u;$$

and by the preceding article,

$$\begin{aligned}\int\delta u = &\int(\mathrm{M}\delta x + \mathrm{M}'\delta dx + \mathrm{M}''\delta d^2x + \&c.)\\ &+\int(\mathrm{N}\delta y + \mathrm{N}'\delta dy + \mathrm{N}''\delta d^2y + \&c.)\end{aligned} \ldots\ldots(1).$$

By the application of the rule for integrating by parts, we find

$$\int \mathrm{M}'\delta dx = \int \mathrm{M}'d\delta x = \mathrm{M}'\delta x - \int d\mathrm{M}'\delta x;$$

$$\begin{aligned}\int \mathrm{M}''\delta d^2x = \int \mathrm{M}''d^2\delta x &= \mathrm{M}''d\delta x - \int d\mathrm{M}''d\delta x\\ &= \mathrm{M}''d\delta x - d\mathrm{M}''\delta x + \int d^2\mathrm{M}''\delta x;\end{aligned}$$

$$\begin{aligned}\int \mathrm{M}'''\delta d^3x = \int \mathrm{M}'''d^3\delta x &= \mathrm{M}'''d^2\delta x - \int d\mathrm{M}'''d^2\delta x\\ &= \mathrm{M}'''d^2\delta x - d\mathrm{M}'''d\delta x + \int d^2\mathrm{M}'''d\delta x\\ &= \mathrm{M}'''d^2\delta x - d\mathrm{M}'''d\delta x + d^2\mathrm{M}'''\delta x - \int d^3\mathrm{M}'''\delta x.\end{aligned}$$

Also

$$\int N'\delta dy = N'\delta y - \int dN'\delta y;$$

$$\int N''\delta d^2y = N''d\delta y - dN''\delta y + \int d^2N''\delta y;$$

$$\int N'''\delta d^3y = N'''d^2\delta y - dN'''d\delta y + d^2N'''\delta y - \int d^3N'''\delta y.$$

Observing that the second member of equation (1) is equal to the sum of the integrals of the terms taken separately, and substituting the above values, we obtain,

$$\begin{aligned}\int\delta u = (M' - dM'' + d^2M''' - \&c.)\,\delta x + (M'' - dM''' + \&c.)\,d\delta x \\ + (M''' - \&c.\ldots\ldots)\,d^2\delta x + \&c. \\ + (N' - dN'' + d^2N''' - \&c.)\delta y + (N'' - dN''' + \&c.)\,d\delta y \\ + (N''' - \&c.\ldots\ldots)\,d^2\delta y + \&c. \\ + \int(M - dM' + d^2M'' - d^3M''' + \&c.)\,\delta x \\ + \int(N - dN' + d^2N'' - d^3N''' + \&c.)\,\delta y. \quad\ldots\ldots\ldots\ldots\ldots(2).\end{aligned}$$

By examining the above expression, it will be seen that there is no term under the sign $\int$ which contains the symbols d and δ applied the one to the other; and also that the parts containing δx are exactly similar to those containing δy. The formula may therefore be extended to any number of variables, by adding for each new variable similar parts containing its variation.

233. It should be remarked, that if the multipliers of δx and δy following the sign $\int$, in equation (2) of the preceding article, are both equal to zero; $\int\delta u$ will be complete, or δu will be the differential of some function. But in the expression

$$\int\delta u = \delta\int u$$

it is evident that if $\int u$ contain any terms which can not be freed from the sign $\int$, $\delta\int u$ must contain the variations of these terms still under the sign, and $\int \delta u$ can not be complete. Hence if δu is a differential, u itself must be so. And conversely; for if $\int u$ is entirely freed from the sign $\int$, then $\delta\int u$ can not contain this sign, and its equal $\int\delta u$ must be complete, or δu be a differential. Hence if the conditions

$$\mathrm{M} - d\mathrm{M}' + d^2\mathrm{M}'' - \&c. = 0$$

$$\mathrm{N} - d\mathrm{N}' + d^2\mathrm{N}'' - \&c. = 0,$$

are satisfied, u will be the differential of some function, which may be obtained by integration.

234. Let us now take the expression $\int v dx$, in which v, as in Art. (231), is a function of x, y, p, q, &c., we have, Arts. (19) and (127),

$$\delta\int v dx = \int\delta(v dx) = \int v\delta dx + \int dx\delta v.$$

But, Art. (140),

$$\int v\delta dx = \int v d\delta x = v\delta x - \int dv\delta x;$$

hence

$$\delta\int v dx = v\delta x + \int(dx\delta v - dv\delta x) \ldots\ldots\ldots\ldots (1).$$

Substituting in that part of the second member which follows the sign $\int$, the values of dv and δv, Arts. (51) and (231),

$$dv = \mathrm{M}dx + \mathrm{N}dy + \mathrm{N}'dp + \mathrm{N}''dq + \&c.;$$

$$\delta v = \mathrm{M}\delta x + \mathrm{N}\delta y + \mathrm{N}'\delta p + \mathrm{N}''\delta q + \&c.;$$

we have

$$dx\delta v - dv\delta x = \mathrm{N}(dx\delta y - dy\delta x) + \mathrm{N}'(dx\delta p - dp\delta x)$$
$$+ \mathrm{N}''(dx\delta q - dq\delta x) + \&c. \ldots\ldots\ldots\ldots (2).$$

Since $dy = pdx$, we have

$$dx\delta y - dy\delta x = dx(\delta y - p\delta x) = \omega dx,$$

by making $\delta y - p\delta x = \omega$.

Also, if for δp we put its value, Art. (231), we have

$$dx\delta p - dp\delta x = d\delta y - pd\delta x - dp\delta x = d(\delta y - p\delta x) = d\omega.$$

If in this last expression we put p for y, and q for p, and recollect that $q = \frac{dp}{dx}$, we have

$$dx\delta q - dq\delta x = d(\delta p - q\delta x) = d\left(\frac{dx\delta p - dp\delta x}{dx}\right) = d\left(\frac{d\omega}{dx}\right).$$

Substituting these values in equation (2), and prefixing the sign $\int$, we have

$$\int(dx\delta v - dv\delta x) = \int \mathrm{N}\omega dx + \int \mathrm{N}'d\omega + \int \mathrm{N}''d\left(\frac{d\omega}{dx}\right) + \&c \ldots\ldots (3).$$

Again by Art. (140),

$$\int \mathrm{N}'d\omega = \mathrm{N}'\omega - \int \frac{d\mathrm{N}'}{dx}\omega dx,$$

$$\int \mathrm{N}''d\frac{d\omega}{dx} = \mathrm{N}''\frac{d\omega}{dx} - \int \frac{d\mathrm{N}''}{dx}d\omega$$

$$= \mathrm{N}''\frac{d\omega}{dx} - \frac{d\mathrm{N}''}{dx}\omega + \int \frac{1}{dx}d\left(\frac{d\mathrm{N}''}{dx}\right)\omega dx.$$

Now substituting these expressions in (3), and the result in (1), we obtain

$$\delta\int v dx = v\delta x + \left(N' - \frac{dN''}{dx} + \&c.\right)\omega + (N'' - \&c.)\frac{d\omega}{dx} + \&c.$$

$$+\int(N - \frac{dN'}{dx} + \frac{1}{dx}d\left(\frac{dN''}{dx}\right) - \&c.)\omega dx.$$

If we now put for ω its value $\delta y - pdx$, the part affected with the sign $\int$ will become

$$\int(N - \frac{dN'}{dx} + \&c.)dx\delta y - \int(N - \frac{dN'}{dx} + \&c.)pdx\delta x.$$

From which we see that, in this case, the coefficients of δx and δy have such a relation that if one becomes equal to zero the other will.

235. The principal, and far the most important application of variations, is to the determination of the *maxima* and *minima* of indeterminate integrals, that is, of integral expressions of the form

$$\int\sqrt{dx^2 + dy^2}, \qquad \int \pi y^2 dx \ldots\ldots \&c.,$$

containing x, y, &c., and their differentials, in which the relation between the variables is entirely unknown. Thus, if it be required to determine the relation between x and y, in order that $\int\pi y^2 dx$ taken under certain conditions, shall be a maximum or minimum, the problem is one not capable of solution by the ordinary method of article (66), since the principles there developed require the form of the function to which they are to be applied, and the constants which enter it, to be given; whereas the object now proposed, is to ascertain what this form and these constants must be, in order that the expression, when subjected to the given conditions, shall be a maximum or minimum. Questions of this kind are readily solved by the aid of variations.

236. Let u be a function of the nature discussed in Art. (231), and suppose x, dx, y, dy, &c., to be increased by their variations, and let the difference between the corresponding function u' and u be developed, which is done at once, by putting δx, δy, δdx, &c., for h, k, l, &c., in the development of Art. (51), we shall thus obtain

$$u' - u = \mathrm{M}\delta x + \mathrm{N}\delta y + \mathrm{M}'\delta dx + \mathrm{N}'\delta dy + \&c.,$$

plus a term of the second degree with respect to δx, δy, &c.; plus other terms.

By the same course of reasoning as that contained in Art. (74), we see that u can be neither greater nor less than u', for all values of δx, δy, &c., unless the term, of the first degree with reference to these variations, is equal to zero. But this term, Art. (231), is the variation of u: *Hence in order that u be a maximum or minimum, δu must be equal to zero.*

If the conditions which make the variation of u equal to zero, make the term of the second degree, in the above development, positive, for all values of δx, δy, &c., u will be a minimum; if negative, u will be a maximum. The discussion of the various circumstances in which this term will not change its sign, is of too complicated a nature, and likely to lead too far, for an elementary treatise. Neither is it necessary, in general, as we shall be able, from the nature of nearly every case, to determine without a reference to this second term, whether we have a maximum or minimum.

237. In the application of the foregoing principles to the indeterminate integrals referred to in Art. (235), it may at first be remarked, that if the integral be *indefinite*, Art. (132), from its nature it can have no maximum nor minimum. The application can then only be made to *definite integrals*, or those which are taken between some well defined limits.

If then, it be required that $\int u$ be a maximum or minimum, we may write the variation of $\int u$, Art. (232), thus,

$$\delta\int u = \int\delta u = m\delta x + n\delta y + m'\delta dx + n'\delta dy + \&c.,$$

$$+ \int(k\delta x + k'\delta y)\ldots\ldots\ldots\ldots(1),$$

and this when taken between the prescribed limits must be equal to zero.

We have seen, Art. (233), that this expression can not be integrated unless the quantity following the sign $\int$ is equal to zero: That is, there can be no integral to be taken between limits, and of course, no maximum nor minimum. We must then have for the first condition

$$k\delta x + k'\delta y = 0\ldots\ldots(2),$$

and since, in general, this must be so for all values of δx and δy, which are independent of each other, we must also have

$$k = 0 \qquad \text{and} \qquad k' = 0$$

or, Art (232),

$$\left.\begin{aligned} M - dM' + d^2M'' - \&c. = 0 \\ N - dN' + d^2N'' - \&c. = 0 \end{aligned}\right\}\ldots\ldots\ldots\ldots(3).$$

Again; if we denote by l and l' the results obtained by substituting the limits in succession, in the remaining part of equation (1), we must have for a second condition,

$$l' - l = 0\ldots\ldots\ldots\ldots(4).$$

Should there be more than two variables in the function u, the quantity following the sign $\int$ in equation (1), will consist of as many terms as there are variables, each of which, if the variations are independent of each other, must be placed equal to zero, and will thus give an equation expressing a relation between these variables and their differentials.

If, however, the conditions under which the variations are made are such as to render these variations in any way dependent, we shall be able, by means of the equations which express these conditions, to eliminate from equation (1), one or more of these variations; then by placing the coefficients of those which remain under the sign $\int$, equal to zero, we shall have a system of equations from which we may determine the nature and extent of the required function. The system of equations (3) will, in every case, express the relation which must exist between the variables and their differentials, in order that the function shall be a maximum or minimum, but they must be subjected to the conditions deduced from the equation

$$l' - l = 0,$$

which can, of course, contain no variables except those which belong exclusively to the limits.

Where u is under the form vdx, it has been seen, Art. (234), that the two equations (3) will both be satisfied, if one is. They will therefore give but one independent equation.

The solution and discussion of the following problems will serve to illustrate and more fully develope the preceding principles.

238. *Problem* 1.—Required the nature of the shortest line joining two given points in a plane.

Let x', y', and x'', y'', be the co-ordinates of the points. The general expression for the length of the line, Art. (197), is

$$z = \int\sqrt{dx^2 + dy^2}.$$

Taking the variation of this, we have

$$\delta\int u = \int\left(\frac{dx\delta dx}{dz} + \frac{dy\delta dy}{dz}\right)$$

which upon comparison with equation (1), Art. (231), gives

$$M = 0, \qquad N = 0, \qquad M' = \frac{dx}{dz}, \qquad N' = \frac{dy}{dz},$$

and all the other terms equal to zero. Hence equations (3), of the preceding article, become

$$d\left(\frac{dx}{dz}\right) = 0 \quad \text{and} \quad d\left(\frac{dy}{dz}\right) = 0,$$

whence by integration,

$$\frac{dx}{dz} = c \qquad \frac{dy}{dz} = c'.$$

Eliminating dz and integrating again, we have

$$dy = \frac{c'}{c}\,dx = adx, \qquad y = ax + b \ldots\ldots(1),$$

which gives the required relation between y and x, and indicates that the line must be straight.

The first part of equation (2), Art. (232), becomes

$$M'\delta x + N'\delta y.$$

Since in this case the limits x' y' and x'' y'' are absolutely fixed, we must have $\delta x'$, $\delta y'$, &c., equal to zero, which being substituted in the above expression give

$$M'\delta x' + N'\delta y' = 0 \qquad M'\delta x'' + N'\delta y'' = 0,$$

whence results the fulfilment of the second condition

$$l' - l = 0,$$

and it remains only to determine the constants a and b, in equation (1), on condition that the line shall pass through the two given points.

239. *Problem* 2. Required the shortest line that can be drawn from one given curve to another.

Let

$$y = f(x) \qquad \text{and} \qquad y = f'(x)$$

be the equations of the curves, their differential equations being

$$dy = p'dx \qquad dy = p''dx......(1).$$

As in the preceding problem, we have

$$z = \int\sqrt{dx^2 + dy^2} \qquad \delta\int u = \int\left(\frac{dx}{dz}\,\delta dx + \frac{dy}{dz}\,\delta dy\right),$$

from which is deduced, precisely as before, the equation of the required line

$$y = ax + b......(2).$$

But since the ends of this line must be in the given curves, the variations of x and y, at the limits, must be confined to these curves, that is, $\delta y'$, $\delta x'$, $\delta y''$, $\delta x''$ must be the same as dy and dx in equations (1), whence

$$\delta y' = p'\delta x' \qquad \delta y'' = p''\delta x''.$$

Substituting these, in succession, in the first part of equation (2), Art. (232), and subtracting the results, we must have

$$l' - l = \left(\frac{dx'}{dz'} + \frac{dy'}{dz'}p'\right)\delta x' - \left(\frac{dx''}{dz''} + \frac{dy''}{dz''}p''\right)\delta x'' = 0,$$

and since this contains two independent variations, it can only be satisfied by making the coefficients separately equal to zero ; hence

$$dx' + dy'p' = 0 \qquad dx'' + dy''p'' = 0,$$

whence

$$\frac{dy'}{dx'} = -\frac{1}{p'} \qquad \frac{dy''}{dx''} = -\frac{1}{p''}.$$

But these are the equations of condition that the required line shall be normal to both curves at the points $(x'\ y')$, $(x''\ y'')$, respectively, Art. (81).

In order to determine the constants a and b in equation (2), we must first find the values of x', y', x'', y'', on condition that the normal to the first curve at the point (x', y') shall also be normal to the second at the point (x'', y''), and then cause the line to pass through these points.

240. *Problem* 3.—Required the shortest line, on the surface of a sphere, joining two given points of the surface.

Let the equation of the sphere be

$$x^2 + y^2 + z^2 = R^2 \ldots\ldots\ldots\ldots(1).$$

The general expression for the length of a line joining the two points will be, Art. (91),

$$w = \int\sqrt{dx^2 + dy^2 + dz^2},$$

the variation of which is

$$\delta \textstyle\int u = \int\left(\frac{dx}{dw}\delta dx + \frac{dy}{dw}\delta dy + \frac{dz}{dw}\delta dz\right);$$

whence, by adding an equation containing δz to those of Art. (233), and comparing, we find

$$M' = \frac{dx}{dw} \qquad N' = \frac{dy}{dw} \qquad P' = \frac{dz}{dw},$$

and thence the first condition required in Art. (237),

$$d\left(\frac{dx}{dw}\right)\delta x + d\left(\frac{dy}{dw}\right)\delta y + d\left(\frac{dz}{dw}\right)\delta z = 0......(2).$$

But in this case the variations must be confined to the surface of the sphere, that is, taking the variation of equation (1), we must have

$$2x\delta x + 2y\delta y + 2z\delta z = 0.$$

Combining this with equation (2), and eliminating δz, we obtain

$$\left\{d\left(\frac{dx}{dw}\right) - d\left(\frac{dz}{dw}\right)\frac{x}{z}\right\}\delta x + \left\{d\left(\frac{dy}{dw}\right) - d\left(\frac{dz}{dw}\right)\frac{y}{z}\right\}\delta y = 0,$$

which, containing two independent variations, gives

$$zd\left(\frac{dx}{dw}\right) - xd\left(\frac{dz}{dw}\right) = 0 \quad zd\left(\frac{dy}{dw}\right) - yd\left(\frac{dz}{dw}\right) = 0.$$

Now if we regard dw as constant, these equations become

$$zd^2x - xd^2z = 0 \qquad zd^2y - yd^2z = 0,$$

from which we deduce

$$xd^2y - yd^2x = 0.$$

Integrating the last three equations, we have

$$zdx - xdz = a, \qquad zdy - ydz = b, \qquad xdy - ydx = c.$$

Multiplying the first by y, the second by $-x$, the third by z, and adding, we obtain

$$ay - bx + cz = 0 \ldots\ldots\ldots\ldots (3),$$

which is the equation of a plane passing through the centre of the sphere. The required curve must lie in this plane, and therefore is the arc of a great circle.

The limits in this case, as in problem 1, being absolutely fixed, we have at once, as in that problem, the fulfilment of the second condition

$$l' - l = 0.$$

Equation (3), may be put under the form

$$\frac{a}{c}y - \frac{b}{c}x + z = 0, \qquad \text{or} \qquad a'y - b'x + z = 0,$$

and the constants a' and b' determined, by causing the plane to pass through the given points.

241. In many cases where there are conditions confining the variations, whether at the limits or not, the method of reducing the number of independent variations explained in Art. (237) and pursued in Arts. (239–40), will be found of very difficult application. In all these cases the following less direct, but very elegant method may be used. Let

$$r = 0 \qquad s = 0 \qquad \&c.$$

be the equations between x, y, &c., expressing the conditions to which the variations are subject; then at the same time that we have

$$\delta \int u = 0,$$

we must also have

$$\delta r = 0 \qquad \delta s = 0, \qquad \&c.,$$

or denoting by c, c', &c., arbitrary constants, we must have the equation

$$\delta \int u + c\delta r + c'\delta s + \&c. = 0 \ldots\ldots (1)$$

for all values of the variations of x, y, &c. Placing the coefficients of these variations separately equal to zero, we obtain equations from which we can eliminate the constants c, c', &c., and thus deduce an equation or equations which will express the proper relation between x, y, &c. As an illustration let us take,

Problem 4.—Required the nature of the line, of a given length, joining two points, which with the ordinates of the points and axis of X, will inclose the greatest area. In this case we have, Art. (203)

$$\delta \int u = \delta \int y dx,$$

and since the length of the arc between the limits is to be constant, the variations must be subject to the condition

$$\int dz = \int \sqrt{dx^2 + dy^2} = a;$$

hence

$$\delta \int \sqrt{dx^2 + dy^2} = 0.$$

Equation (1) will then become

$$\delta\int y dx + c\delta\int\sqrt{dx^2 + dy^2} = 0,$$

or putting for the variations their values, we have

$$\int\left(y\delta dx + dx\delta y + \frac{cdx\delta dx + cdy\delta dy}{dz}\right) = 0.$$

Comparing this with equation (1), Art. (231), we see that

$$M = 0, \quad M' = y + c\frac{dx}{dz}, \quad N = dx, \quad N' = c\frac{dy}{dz},$$

and these being substituted in equations (3), of Art. (237), give

$$-d\left(y + c\frac{dx}{dz}\right) = 0 \qquad dx - cd\left(\frac{dy}{dz}\right) = 0,$$

and by integrating

$$y + c\frac{dx}{dz} = b \qquad x - c\frac{dy}{dz} = b'.$$

Eliminating c from these two equations, we obtain

$$\frac{dy}{dx} = -\frac{x - b'}{y - b},$$

which is evidently the differential equation of a circle whose equation is, (Art. 98),

$$(y - b)^2 + (x - b')^2 = R^2,$$

b, b' and R being arbitrary constants, which must be determined on condition that the circle pass through the two given points, and that the included arc be of the given length.

www.ingramcontent.com/pod-product-compliance
Lightning Source LLC
LaVergne TN
LVHW010153110826
845151LV00002B/504

* 9 7 8 1 4 2 5 5 3 7 0 1 2 *